Atomic and Molecular Manipulation

FRONTIERS OF NANOSCIENCE

Series Editor: Richard E. Palmer
The Nanoscale Physics Research Laboratory,
The School of Physics and Astronomy,
The University of Birmingham, UK

Vol. 1 **Nanostructured Materials** edited by Gerhard Wilde

Vol. 2 **Atomic and Molecular Manipulation** edited by Andrew J. Mayne and Gérald Dujardin

Atomic and Molecular Manipulation

Edited by

Andrew J. Mayne and Gérald Dujardin
Institut des Sciences Moléculaires d'Orsay
CNRS, UMR 8214, Bâtiment 210
Université de Paris-Sud 11
Orsay F-91405, France

ELSEVIER
Amsterdam • Boston • Heidelberg • London • New York • Oxford
Paris • San Diego • San Franciso • Singapore • Sydney • Tokyo

Elsevier
The Boulevard, Langford Lane, Kidlington, Oxford OX5 1GB, UK
Radarweg 29, PO Box 211, 1000 AE Amsterdam, The Netherlands

British Library Cataloguing in Publication Data
A catalogue record for this book is available from the British Library

Library of Congress Cataloging-in-Publication Data
A catalog record for this book is available from the Library of Congress

ISSN: 1876-2778
ISBN: 978-0-08-096355-6

Printed and bound in Great Britain

11 12 13 14 10 9 8 7 6 5 4 3 2 1

Contents

Contributors ix

1. Introduction 1

Andrew J. Mayne and Gérald Dujardin

Part I
Electronics at the Atomic-Scale 15

2. STM Manipulation of Single Atoms and Molecules on Insulating Films 17

Jascha Repp and Gerhard Meyer

1. Introduction 17
2. Ultrathin Insulating Films 18
3. Interface State in NaCl/Cu(111) 22
4. Manipulation of Metal Atoms 24
4.1. Controlling the Charge State 24
4.2. Vertical Transfer and Build-up of Nanostructures 28
5. Imaging Molecular Orbitals 30
6. Manipulation of Molecules: Molecular/Orbital Engineering 35
6.1. Manipulation of a Small Molecule: Water Dissociation on MgO Films 36
6.2. Organometallic Synthesis: Au–Pentacene 36
6.3. Metal–Ligand Complex Formation 40
6.4. Molecular Switching Based on a Tautomerization Reaction 41
7. Conclusion 44
References 45

3. Electron Transfer Phenomena at the Molecular Scale: Organic Charge Transfer Complexes on Metal Surfaces 51

Isabel Fernández Torrente, Katharina J. Franke and Jose Ignacio Pascual

1. Introduction 51
2. TCNQ on Au(111): The Neutral Adsorption of an Electron Acceptor Molecule 54
3. TTF on Au(111): Formation of a Wigner Molecular Lattice 57
3.1. TTF Nucleation: Long-Range Repulsive Versus Short-Range Attractive Interactions 64

4. TTF–TCNQ on Au(111): Molecular Magnetism Induced by CT 65
5. Conclusions and Remarks 71
References 72

Part II
Chemistry at the Atomic-Scale 77

4. Imprinting Atomic and Molecular Patterns 79

Iain R. McNab and John C. Polanyi

1. Single-Molecule Imprinting on Semiconductors 79
1.1. Localized Atomic Reaction 81
1.2. Single Dissociative Attachment 1: Diatomic Molecules at Si(100)-2 × 1 84
1.3. Single Dissociative Attachment 2: Dissociative Attachment of Polyatomics 86
1.4. Discussion of Single Dissociative Attachment 91
1.5. Multiple Dissociative Attachment 91
2. Single-Molecule Imprinting on Metals 94
2.1. LAR on Metals 94
3. Self-Assembly Followed by Pattern Imprinting 98
3.1. Imprinting Circles 99
3.2. Forming Lines 103
3.3. Imprinting Lines 103
4. Modes of Reaction 106
4.1. Direct and Indirect Reaction 106
4.2. Cooperative Reaction 107
4.3. Chain Reaction 108
4.4. Recoil Reaction 109
5. Conclusion 113
Acknowledgements 114
References 114

5. Tunnel-Current Induced STM Atomic Manipulation 121

Peter A. Sloan

1. Introduction 121
2. Experimental Methods 122
2.1. Frame-by-Frame Scanning 122
2.2. Single-Point Current Injection 122
3. Desorption 123
3.1. DIET and DIMET 124
3.2. Si(100)-2 × 1:H 126
3.3. Chlorobenzene/Si(111)-7 × 7 127
4. Intramolecular Bond Dissociation 131
4.1. O_2/Pt(111) 132
4.2. Chlorobenzene/Si(111)-7 × 7 134

4.3. Fluoropentane/Si(100) 137
5. Nonlocal Manipulation 138
5.1. CH_3SSCH_3/Au(111) 139
5.2. Chlorobenzene/Si(111)-7 × 7 141
References 145

Index 151

Colour versions of figures in this book can be found at http://www.elsevierdirect.com/product.jsp?isbn=9780080963556

Contributors

Numbers in Parentheses indicate the pages on which the author's contributions begin.

Gérald Dujardin (1), Institut des Sciences Moléculaires d'Orsay, CNRS, Université Paris Sud 11, Orsay, France

Katharina J. Franke (51), Institute für Experimentalphysik, Freie Universität Berlin, Arnimallee, Berlin, Germany

Andrew J. Mayne (1), Institut des Sciences Moléculaires d'Orsay, CNRS, Université Paris Sud 11, Orsay, France

Iain R. McNab (79), Department of Chemistry and Institute of Optical Sciences, University of Toronto, Toronto, Ontario, M5S 3H6, Canada

Gerhard Meyer (17), IBM Research-Zurich, Rüschlikon, Germany

Jose Ignacio Pascual (51), Institute für Experimentalphysik, Freie Universität Berlin, Arnimallee, Berlin, Germany

John C. Polanyi (79), Department of Chemistry and Institute of Optical Sciences, University of Toronto, Toronto, Ontario, M5S 3H6, Canada

Jascha Repp (17), Institute of Experimental and Applied Physics, Universität Regensburg, Germany

Peter A. Sloan (121), Department of Physics, University of Bath, Bath, United Kingdom

Isabel Fernández Torrente (51), Institute für Experimentalphysik, Freie Universität Berlin, Arnimallee, Berlin, Germany

Chapter 1

Introduction

Andrew J. Mayne and Gérald Dujardin
Institut des Sciences Moléculaires d'Orsay, CNRS, Université Paris Sud 11, Orsay, France

Working with individual atoms and molecules is a demonstration that miniaturised electronic, optical, magnetic and mechanical devices can operate ultimately even at the level of a single atom or molecule. As such, manipulation of individual atoms and molecules with the scanning tunnelling microscope (STM) has played a very emblematic role in the development of nanosciences and nanotechnologies. New methods, based on the use of the STM, have been developed to characterise and manipulate all the degrees of freedom of individual atoms and molecules, electronic, vibrational, spin spectroscopy, lateral and vertical manipulation,[1] with an unprecedented precision. Based on this knowledge, manipulation of individual atoms and molecules has been used in recent years to test and propose new concepts for nanoelectronics, molecular nanomachines and the design of functionalised materials. From this, it appears that manipulation of individual atoms and molecules goes well beyond the demonstration of spectacular experiments and is now able to provide innovative concepts. This book illustrates the main aspects of this ongoing scientific adventure and anticipates the major challenges for the future in 'atomic and molecular manipulation' from fundamental knowledge to the fabrication of new devices and materials.

This book is divided into two sections which illustrate some of the different research directions in atomic and molecular manipulation on surfaces:

1. Electronics at the atomic scale
   Jascha Repp and Gerhard Meyer, 'STM manipulation of single atoms and molecules on insulating films'
   Isabel Fernandez Torrente, Katharina J. Franke and Jose I. Pascual, 'Electron transfer phenomena at the molecular scale: Organic charge transfer complexes on metal surfaces'
2. Chemistry at the atomic scale
   John C. Polanyi and Iain R. McNab, 'Imprinting atomic and molecular patterns'
   Peter A. Sloan, 'Tunnel current induced STM atomic manipulation'

Frontiers of Nanoscience, Vol. 2. DOI: 10.1016/B978-0-08-096355-6.00001-5

Nanoscience really entered the public consciousness in 1990 with the writing of the IBM logo in Xenon atoms by Eigler.[2] This clearly showed that it was possible, using the STM, to do experiments and to fabricate things with only a few atoms and molecules and ultimately a single atom or molecule. Since 1990, there have been spectacular demonstrations of single atom or molecular manipulation. These range from the manipulation of a wide range of individual atoms and molecules to far more sophisticated experiments where complex nano-objects are fabricated by STM manipulation.[1,3,4]

The basic design and operation of the STM have been explained in earlier books or reviews.[5,6] Low temperature operation and sophisticated software have improved STM performances over the years.[7] Further recent developments are discussed at the end of this introduction. Manipulation methods in STM can be categorised into two principal types, manipulation by direct contact and manipulation by inelastic electron tunneling.[3,4]

The direct contact method involves to bring the tip into close proximity with the adsorbate and to induce a modification by moving the tip. This can be either lateral, that is, parallel to the surface, or vertical, that is, normal to the surface. In these manipulations, the active ingredient is neither the tunnel electrons nor the electric field but rather the forces between the atoms on the tip and the surface. Historically, the direct contact method arose from the fact that the early manipulation experiments, by Eigler, for example, were carried out on metal surfaces where manipulation induced movement in the surface plane. Lateral manipulation is most widely used on metal surfaces because metals have a relatively uniform surface potential allowing adsorbed molecules to be pushed, pulled or slid across the surface.[8] Over the years, a number of spectacular results have been produced. Metal atoms can be positioned individually on a metal surface to create different geometric shapes, such as circles, ellipses, squares or resonators. These quantum corrals act as electron traps with the formation of standing waves of electronic surface states inside the corral.[9,10] One of the most spectacular atomic-scale experiments is the quantum mirage where the placement of a magnetic Cobalt atom at a focus of an elliptical corral led to the appearance of a phantom atom at the other focus (Figure 1.1).[11] Further experiments engineered different atomic structures using the coupling of the spin of magnetic atoms.[12] Atom manipulation on metal surfaces has become a fine art, whereby judicious positioning of metal atoms can control the confinement of quantum interference patterns to the extent that the metal can be quantum holographically encoded (Figure 1.2).[13] The direct contact between the STM tip and the surface can also be used for vertical manipulation of single atoms or molecules. This has been demonstrated by the vertical manipulation of individual germanium atoms from a Ge(111)-$c(2\times 8)$ surface at room temperature.[14,15]

Inelastic electron tunnelling has been proved to be the most versatile method for STM manipulation.[1,3,4] Of special interest is the resonant inelastic electron tunnelling when tunnel electrons are in resonance with occupied or

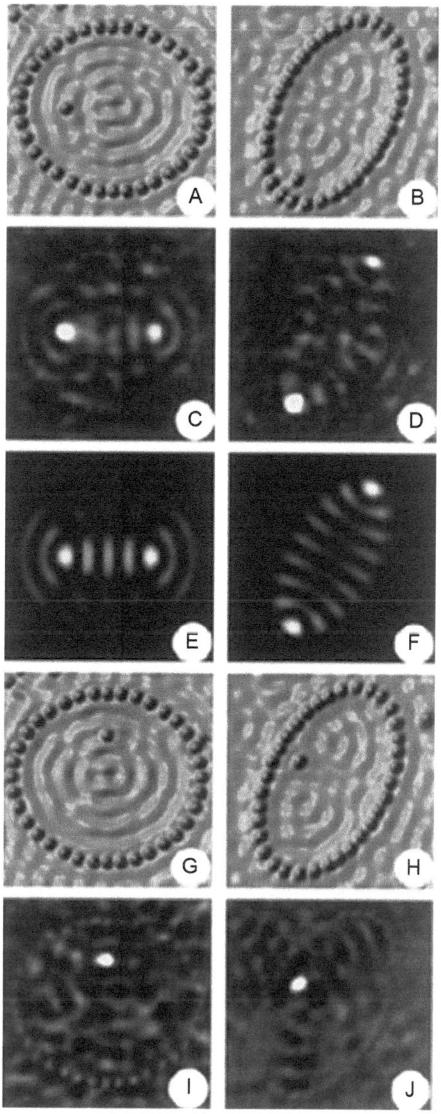

FIGURE 1.1 Visualisation of the quantum mirage. (A, B) Topographs showing the $e=1/2$ (A) and $e=0.786$ (B) ellipse each with a Co atom at the left focus. (C, D) Associated dI/dV difference maps showing the Kondo effect projected to the empty right focus, resulting in a Co atom mirage. (E and F) Calculated eigenmodes at E_F (magnitude of the wavefunction is plotted). When the interior Co atom is moved off focus (g and h, topographs), the mirage vanishes (i and j, corresponding dI/dV difference maps). Imaging conditions and dimensions ($V-10$ mV, $V_{a.c.}=250$ μV r.m.s; $I=1$ nA for A, C, G–I; $V=8$ mV, $V_{a.c.}=1$ mV r.m.s., $I=1$ nA for B, D, H and I). Image dimensions are 150 Å square and 154 Å square for the $e=1/2$ and $e=0.786$ ellipses, respectively. We have assembled over 20 elliptical resonators of varying size and eccentricity and searched for the formation of a quantum mirage. We find that as a is increased monotonically while e is fixed, the mirage is switched on and off. In each period of this switching, the classical path length $2a$ changes by a half Fermi wavelength. Because we also observe that two focal atoms, one on each focus, couple quite strongly with one another (as judged by the perturbation of the Kondo resonance), these oscillatory results may have a source akin to the RKKY (indirect exchange) interaction.

unoccupied states of adsorbed atoms and molecules. Most of the time, resonant inelastic electron tunnelling results in local vibrational excitation, which can be used to desorb,[16] to move single atoms and molecules or to dissociate molecules,[17] and isomerise or change the configuration of a single molecule.[18–23] To increase the efficiency of resonant inelastic electron tunnelling, molecules should ideally be decoupled electronically from the substrate. So a number of avenues of research have investigated different approaches towards electronic decoupling of molecular orbitals from the underlying surface.[24]

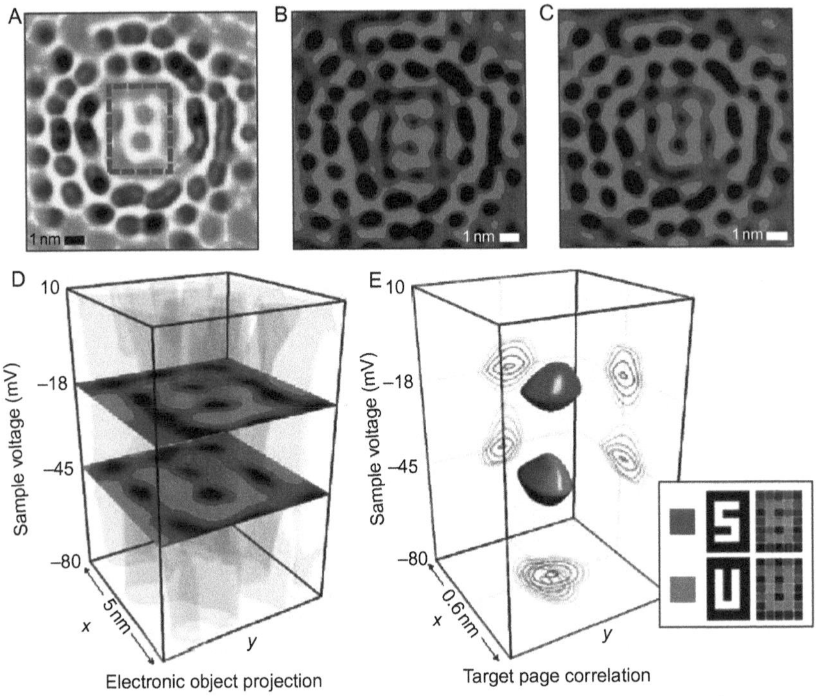

FIGURE 1.2 Volumetric quantum holography. (A) Topograph (13.5×13.5 nm^2; $I=1$ nA; $V=260$ mV) of a molecular hologram that encodes two pages of data at different energies in the same region of space. (B) A dI/dV map taken at $V=218$ mV shows the S page. $V_{a.c.}=4$ mV r.m.s. (C) Measuring dI/dV at $V=245$ mV retrieves the U page. The molecules are unchanged. (D) By mapping dI/dV at many voltages between 280 and 10 mV, $\rho(r, E)$ was measured throughout the readout region. A translucent surface of constant DOS is shown. Slices through this space at the appropriate energies reveal the S and U pages. (E) The normalised cross-correlation of $\rho(r, E)$ with each template image (insets) was computed as a function of r and E. Surfaces of constant correlation (at 98% of the global maximum) are shown for each page, confirming their locations in the information cube. Contours show the maximum correlation projected in each dimension (95–99.9%).

Molecular deposition and manipulation have been tested on several different surfaces; thin insulating layers on a metal substrate,[25–27] passivated semiconductor surfaces[28–31] and bulk insulator surfaces.[32,33] As a general rule, in the case of these three categories of insulating surfaces, the molecules are only very weakly bound. To immobilise the molecules and so avoid surface diffusion on surfaces, passivated or otherwise, there are several options; use low temperature studies, self-assembled molecular arrays,[34–36] or molecules with organic ligands to increase the binding energy.[37,38] An alternative avenue of research is a wide band gap semiconductor such as silicon carbide. Until recently, most of the focus had been on the structural and electronic properties of silicon carbide,[39–41] but its reactivity towards molecules had received very little attention, with experiments only of oxidation[42,43] or hydrogenation.[44] Theoretical studies predicting reaction with organic molecules prompted recent STM studies of the adsorption of large organic molecules.[45] These new STM and STS results show that the wide band gap of silicon carbide can indeed be used to electronically decouple the molecules from the surface.[46,47]

The four chapters in this book are not limited to the description and the use of the various methods for STM manipulation. Each chapter illustrates how atomic and molecular manipulation has opened up completely new concepts for nanoelectronics (Chapters 2 and 3) and the design of new functionalised materials (Chapters 4 and 5). Complementary calculations have now become *de rigueur* for almost every STM experiment, be that a simple observation to complex molecular dynamics during STM manipulation.[48] A variety of calculation methods are used from *ab initio* and semi-empirical calculations through to simulated STM images of the atoms or molecules adsorbed on a surface.

In Chapter 2, Jascha Repp and Gerhard Meyer review their recent approach of using thin insulating layers on metal surfaces for manipulating single atoms and molecules. Clean metal surfaces, which have been initially used for STM manipulation, may have some advantages for lateral manipulation due to low diffusion barriers of adsorbed atoms and molecules. However, they are obviously not ideally suited to investigate electronic properties of single atoms and molecules whose orbitals strongly couple with the surface states of metals. The role of a thin insulating layer deposited on a metal substrate is to electronically decouple adsorbed atoms and molecules from the substrate by introducing a second tunnel barrier between the atoms and molecules and the metal substrate. By this method, Repp and Meyer are able to considerably extend the capabilities of single-atom and single-molecule electronics. For example, they demonstrate the control of charge states of single atoms (Au, Ag). Due to the large ionic polarisability of the thin insulating film of alkali halides (NaCl), several charge states of the Au or Ag atoms are stable. Then the STM tip is used to charge and discharge at will a single atom with a unity quantum yield, that is, a single tunnel electron is sufficient. Such a giant efficiency of resonant inelastic electron tunnelling (usually the quantum yield is between 10^{-6} and 10^{-10}) indicates that the resonance is very long lived, thus confirming the insulating role of the thin NaCl film. This ability to manipulate a single electron at the level of a single atom opens the perspectives of future single-electron and single-atom electronics. Repp and Meyer introduce another very promising concept, that of molecular orbital engineering. They show that single-molecular orbitals can be directly imaged with the STM when the molecule is decoupled from the metal substrate by the insulating thin film. This ability to monitor a single-molecular orbital and its detailed modifications upon STM manipulation is exploited to perform experiments that would be impossible on metal substrates. By various sequences of STM manipulations, Repp and Meyer are able to activate an organometallic synthesis (bond formation between a single metal atom (Au) and a molecule (pentacene)) or a coordination complex synthesis (bond formation between a single Fe atom and two molecules). Single molecular orbital imaging enables an unambiguous control of the synthesis results. This is also used to control a molecular switching based on a tautomerisation reaction induced by inelastic electron tunnelling with the STM. These results illustrate the huge potential of molecular orbital engineering to monitor in detail all kinds of molecular reactions.

In Chapter 3, Isabel Torrente, Katharina Frenke and Jose Pascual report their discovery of new types of materials, that is, metal-free molecular magnets. This is very emblematic of the new knowledge afforded by STM topography and spectroscopy at the molecular scale. Here, the authors' objectives are to understand charge-transfer processes in organic charge-transfer materials. In crystal form, these compounds form semiconducting organic solids which are attractive for applications in organic electronics and photovoltaics. The particular molecules chosen here are the well-known tetrathiafulvalene (TTF) and tetracyanoquinodimethane (TCNQ) which have the ability to either donate (TTF) or accept electrons (TCNQ).[49] For their STM studies, Torrente, Frenke and Pascual produce a self-assembled layer of TTF and TCNQ molecules on a metal surface (Au(111)). By carefully studying low temperature local STM spectroscopy and microscopy, the authors resolve the electron redistribution processes taking place at the organic-metal interface. The outstanding result is that the TCNQ molecule is charged with one electron. This negatively charged molecular radical thus sits in a spin ½ state and becomes magnetic. This is evidenced by a Kondo resonance[50] analysed by STM spectroscopy which shows vibrational fingerprints. From these results, the authors conclude that organic donor–acceptor interactions are promising avenue towards the spontaneous self-assembled organisation of metal-free molecular magnets.

Chapter 4 by Iain McNab and John Polanyi discusses the concept of molecular scale imprinting (MSI). It entails the self-assembling of atomic and molecular patterns on semiconductor or metal surfaces followed by localised atomic reactions that can be activated either locally by the tunnelling current from an STM tip or by heating or by photon irradiation. These two concepts of MSI and localised atomic reactions introduced by the Canadian group are issued from very detailed studies of molecular reactions on surfaces at the atomic scale with the STM.[51,52] Many types of molecular reactions are discussed in this chapter: single and double dissociative attachment reactions, cooperative reactions, chain reactions, recoil reactions. Also, self-assembling of molecules into various patterns, circles and lines are shown. These examples illustrate the unique advantage of STM to explore the complexity of molecular reactions on surfaces. From this knowledge, McNab and Polanyi are able to extract new methods for imprinting atomic and molecular patterns.

In Chapter 5, Peter Sloan reviews some of the fundamental processes of STM molecular manipulation, desorption and intramolecular bond dissociation, produced by inelastic electron tunnelling. These processes are key processes in catalysis or surface photochemistry. They also provide new insights in electron-induced reactions in semiconductors or condensed gases. This chapter brings us up to date by describing the subtle interplay between competing dissociation and desorption reaction pathways of molecules that can be tuned by suitable adjustment of the electron tunnel conditions.[53,54] So far, most of the studies have considered the surface as a passive substrate. Peter Sloan shows, from recent examples, that non-local reactions involving the transfer of energy or charges across surface states can also produce reactions at a distance.

This book is focused on STM manipulation of atoms and molecules and yet several chapters allude to wider possibilities afforded by atomic force microscopy (AFM) in the near future. We will end this introduction with a glimpse into what AFM might be able to accomplish. AFM is widely used nowadays as a surface probe in many fields of research and industry, provided only nanometer resolution is required. However, AFM has so far proved to be a more complicated technique, and so obtaining atomic resolution has been a long hard road. Nevertheless, in recent years since the development of the tuning-fork cantilever,[55] much progress has been made in using non-contact UHV AFM. To illustrate this, we can look at a couple of recent studies. The AFM can be used to manipulate single atoms across surfaces using lateral manipulation where the tip does not actively participate in the manipulation—it does not pick up any atoms.[56] The Sn/Si(111) surface provides an interesting substrate in that Si atoms can populate the Sn adlayer. They show up as slightly darker 'defect' atoms. It should be remembered that imaging atoms with the AFM is not simple and that contrast is due to differences in the interaction between the tip and the atom on the surface.[57,58] The force acting between the AFM tip and the atom can be used to manipulate single atoms or molecules, for example, CO on copper[59] or Sn atoms on Ge(111)-*c* (2×8).[56] Both these results were based on lateral manipulation of the adsorbed species. However, recent AFM results show (Figure 1.3) that, in vertical manipulation, the tip plays an active role by exchanging the surface atom with one on the tip.[60] Earlier STM experiments on the Ge(111) surface had shown that the STM tip, when placed in contact with an atom on the substrate, could remove a single atom by forming a chemical bond with the tip.[15,61,62] In these studies, single atom removal is perfectly controlled which led to studies of surface diffusion,[61,63] reactivity[64] and the role in the initial stages of the oxidation of the Ge(111) surface.[65]

Recent studies with the AFM have taken a very important step forward: single chemical bonds of a pentacene molecule have been imaged with unprecedented resolution (Figure 1.4).[66] Using this new capability, the organic structure of the cephalandole molecule which could not be determined by any

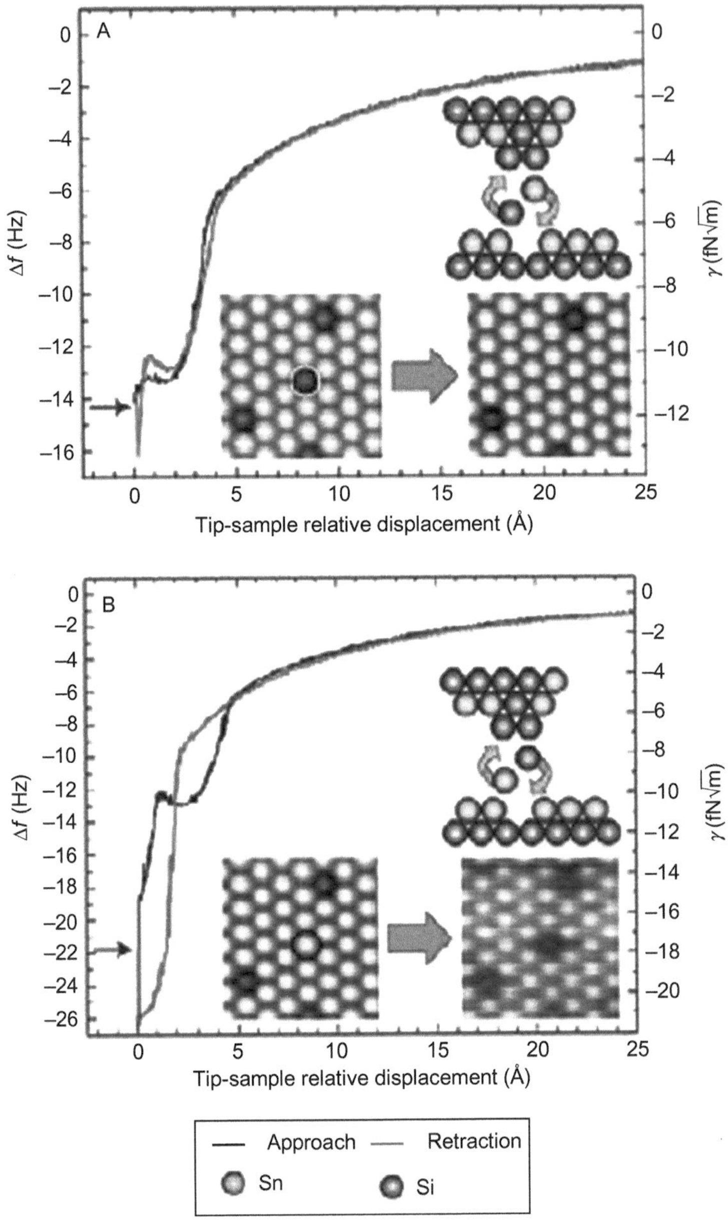

FIGURE 1.3 Alternate vertical interchange atom manipulations. (A) Frequency shift (Δf) signal upon approach (black) and retraction (red) of the tip over the Si atom marked with a white circle in the left inset image. In a consecutive topographic image to the curve acquisition (bottom right inset), an Sn atom appears at the same surface location instead. The Si atom was replaced by an Sn atom coming from the tip (Sn deposition). (B) Frequency shift signal upon approach (black) and retraction (red) of the tip above the Sn atom deposited in (A), pointed out by a black circle

other technique has been elucidated using atomic resolution AFM.[67] Further, NC–AFM has been used to measure the charge state of an adatom,[68] which opens up new perspectives in molecular electronics, photonics, catalysis and solar photo-conversion. These results clearly indicate the extent to which AFM is now following in the footsteps of the STM, especially considering that AFM can be used on almost any surface, in particular, bulk insulators such as CaF_2,[69] MgO[70] and diamond.[71] In addition, the AFM can manipulate large nano-objects such as CdSe nanorods[72] as well as individual atoms.

Recent instrumental developments in scanning tunnelling microscopy enable very low temperature (down to 10 mK)[73] and high magnetic field operation, either externally or locally in twisted nanostructures.[74] This will allow the manipulation of new quantum states of matter (e.g. superconductivity and competing or hidden

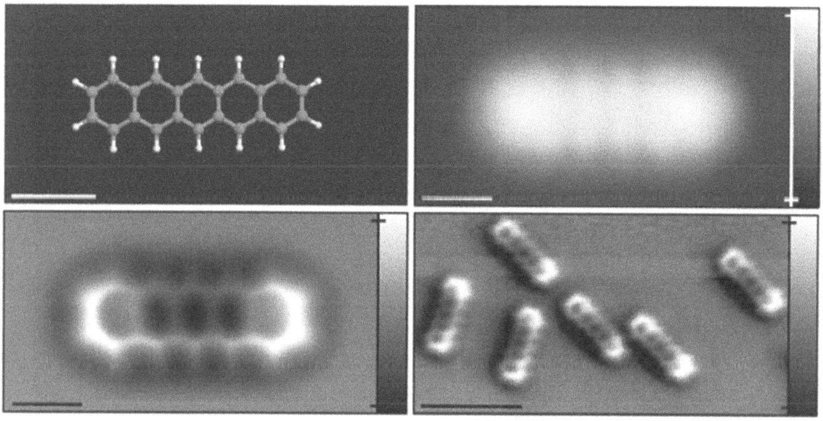

FIGURE 1.4 STM and AFM imaging of pentacene on Cu(111). (A) Ball-and-stick model of the pentacene molecule. (B) Constant-current STM and (C and D) constant-height AFM images of pentacene acquired with a CO-modified tip. Imaging parameters are as follows: (B) set point $I = 110$ pA, $V = 170$ mV; (C) tip height $z = -0.1$ Å (with respect to the STM set point above Cu (111)), oscillation amplitude $A = 0.2$ Å; and (D) $z = 0.0$ Å, $A = 0.8$ Å. The asymmetry in the molecular imaging in (D) (showing a 'shadow' only on the left side of the molecules) is probably caused by asymmetric adsorption geometry of the CO molecule at the tip apex.

(left inset). After the curve acquisition, the replacement of this Sn atom by an Si atom coming from the tip (Si deposition) and a partial loss of atomic contrast are obtained (bottom right inset). For comparison with other curves, the normalised frequency shift (γ) is displayed in the vertical axis on the right. The black arrows in the plots indicate instabilities representative of the corresponding concerted vertical interchange of atoms between the tip and surface. The origin of the horizontal axes denotes the point of maximum proximity between the tip and sample; this criterion was adopted for all the experimental curves shown in this work. The illustrations are representations of the corresponding vertical atomic interchange, with yellow and grey spheres symbolising Sn and Si atoms, respectively; they do not bear any realistic information about the tip apex structure or composition. (For interpretation of the references to colour in this figure legend, the reader is referred to the Web version of this chapter.)

states, charge ordering[75]) as well as single spin detection in individual atoms[76] and molecules[77] and the fabrication of atomic structures that could behave as quantum dots.[78–80] Atomic and molecular manipulation is also making progress, thanks to the development of new materials which can be studied at the atomic scale; new substrate materials include magnetic semiconductors,[81] graphene[82–84] and silicene.[85,86] Further, the newly developed low temperature (4 K) STM multiprobe machines with four fully independent probes[87] will open new avenues of research in quantum and molecular electronics.

To conclude, this book has been written to illustrate how STM manipulation has opened new areas of research in molecular electronics, molecular nanomachines, catalysis, etc.[88–90] From this, the development of a real atom-scale technology is emerging. STM manipulation of atoms and molecules has evolved step by step, with each step opening up completely new perspectives that were impossible to anticipate at any stage before. The visualisation of manipulating the building blocks of matter, taken in the larger context of nanotechnology, has generated both fascination and curiosity from the general public. This is perhaps, in part, due to the parallel progression of scientific discovery enabled by the STM and AFM along with the technological progress in everyday life blurring our fundamental conceptions of nature and artificial.[91] Hopefully, the exciting discoveries in STM manipulation of atoms and molecules will act as a very convincing illustration teaching us how to advance science at the beginning of the twenty-first century.

REFERENCES

1. Mayne AJ, Dujardin G, Comtet G, Riedel D. Electronic control of single-molecule dynamics. *Chem Rev* 2006;**106**:4355–78.
2. Eigler DM, Schweizer EK. Positioning single atoms with a scanning tunnelling microscope. *Nature* 1990;**344**:524.
3. Mayne AJ, Dujardin G. STM manipulation and dynamics. In: Hasselbrink E, Lundqvist BI, editors. *Handbook of surface science*, vol. 3. North Holland: Elsevier; 2008. p. 681–759.
4. Mayne AJ, Riedel D, Comtet G, Dujardin G. Electronic control of single molecule nanomachines. In: Seideman T, editor. *Current-driven phenomena in nanoelectronics*. Singapore: Pan Scientific Publishing; 2010. 978-981-4241-50-2.
5. Stroscio JA, Feenstra RM. Stroscio JA, Kaiser WJ, editors. *Scanning tunneling microscopy, methods of experimental physics*. New York: Academic Press; 1993.
6. Kubby JA, Boland JJ. Scanning tunnelling microscopy of semiconductor surfaces. *Surf Sci Rep* 1996;**26**:61.
7. Meyer G, Repp J, Zöphel S, Braun KF, Hla SW, Fölsch S, Bartels L, Moresco F, Rieder KH. Controlled manipulation of atoms and small molecules with a low temperature scanning tunneling microscope. *Single Mol* 2000;**1**:79–86.
8. Bartels L, Meyer G, Rieder K-H. Basic steps of lateral manipulation of single atoms and diatomic clusters with a scanning tunneling microscope tip. *Phys Rev Lett* 1997;**79**:697.
9. Crommie MF, Lutz CP, Eigler DM. Confinement of electrons to quantum corrals on a metal surface. *Science* 1993;**262**:218.

10. Kliewer J, Berndt R, Crampin S. Scanning tunnelling spectroscopy of electron resonators. *New J Phys* 2001;**3**:22.
11. Manoharan HC, Lutz CP, Eigler DM. Quantum mirages formed by coherent projection of electronic structure. *Nature* 2000;**403**:512.
12. Hirjibehedin CF, Lutz CP, Heinrich AJ. Spin coupling in engineered atomic structures. *Science* 2006;**312**:1021.
13. Moon CR, Mattos LS, Foster BK, Zeltzer G, Manoharan HC. Quantum holographic encoding in a two-dimensional electron gas. *Nature Nanotechnology* 2009;**4**:167.
14. Becker RS, Golovchenko JA, Swartzentruber BS. Atomic-scale surface modifications using a tunnelling microscope. *Nature* 1987;**325**:419.
15. Dujardin G, Mayne AJ, Rose F, Robert O, Tang H, Joachim C. Vertical manipulation of individual atoms by a direct STM tip-surface contact on Ge(111). *Phys Rev Lett* 1998;**80**:3085.
16. Bartels L, Wolf M, Klamroth T, Saalfrank P, Kühnle A, Meyer G, et al. Atomic-scale chemistry: Desorption of ammonia from Cu(111) induced by tunneling electrons. *Chem Phys Lett* 1999;**313**:544.
17. Dujardin G, Walkup RE, Avouris Ph. Dissociation of individual molecules with electrons from the tip of a scanning tunneling microscope. *Science* 1992;**255**:1232.
18. Stipe BC, Rezaei M, Ho W. Single-molecule vibrational spectroscopy and microscopy. *Science* 1998;**280**:1732.
19. Stipe BC, Rezaei MA, Ho W. Coupling of vibrational excitation to the rotational motion of a single adsorbed molecule. *Phys Rev Lett* 1998;**81**:1263.
20. Dujardin G, Rose F, Mayne AJ. Toggling the local surface work function by pinning individual promoter atoms. *Phys Rev B* 2001;**63**:235414.
21. Martin M, Lastapis M, Riedel D, Dujardin G. Mastering the molecular dynamics of a bistable molecule by single atom manipulation. *Phys Rev Lett* 2006;**97**:216103.
22. Mayne AJ, Lastapis M, Baffou G, Soukiassian L, Comtet G, Hellner L, et al. A chemisorbed bistable molecule: biphenyl on Si(100)-2 × 1. *Phys Rev B* 2004;**69**:045409.
23. Cranney M, Mayne AJ, Comtet G, Dujardin G. STM excitation of individual biphenyl molecules: DIET or DIEF? *Surf Sci* 2005;**593**:139.
24. Mayne AJ, Riedel D, Comtet G, Dujardin G. Atomic-scale studies of hydrogenated semiconductor surfaces. *Prog Surf Sci* 2006;**81**:1–52.
25. Qiu XH, Nazin GV, Ho W. Vibrationally resolved fluorescence excited with submolecular precision. *Science* 2003;**299**:542–6.
26. Liljeroth P, Repp J, Meyer G. Current-induced hydrogen tautomerization and conductance switching of naphthalocyanine molecules. *Science* 2007;**317**:1203–6.
27. Villagomez CJ, Zambelli T, Gauthier S, Gourdon A, Stojkovic S, Joachim C. STM images of a large organic molecule adsorbed on a bare metal substrate or on a thin insulating layer: visualization of HOMO and LUMO. *Surf Sci* 2009;**603**:1526–32.
28. Mayne AJ, Soukiassian L, Commaux N, Comtet G, Dujardin G. Molecular molds. *Appl Phys Lett* 2004;**85**:5379–81.
29. Soukiassian L, Mayne AJ, Carbone M, Dujardin G. Atomic wire fabrication by STM induced hydrogen desorption. *Surf Sci* 2003;**528**:121–7.
30. Bellec A, Ample F, Riedel D, Dujardin G, Joachim C. Imaging molecular orbitals by scanning tunneling microscopy on a passivated semiconductor. *Nano Lett* 2009;**9**:144–7.
31. Dujardin G, Mayne AJ, Rose F. Temperature control of electronic channels through a single atom. *Phys Rev Lett* 2002;**89**:036802.
32. Schülte J, Bechstein R, Rohlfing M, Reichling M, Kühnle A. Cooperative mechanism for anchoring highly polar molecules at an ionic surface. *Phys Rev B* 2009;**80**:205421.

33. Tekiel A, Godlewski S, Buzioch J, Szymonski M. Nanofabrication of PTCDA molecular chains on rutile $TiO_2(001)$-(2×1) surfaces. *Nanotechnology* 2008;**19**:495304.
34. Swarbrick JC, Ma J, Theobald JA, Oxtoby NS, O'Shea JN, Champness NR, et al. Square, hexagonal, and row phases of PTCDA and PTCDI on Ag-Si(111) $\sqrt{3} \times \sqrt{3}R30°$. *J Phys Chem B* 2005;**109**:12167–74.
35. Ramoino L, Von Aix M, Schintke S, Baratoff A, Güntherodt H-J, Jung TA. Layer-selective epitaxial self-assembly of porphyrins on ultrathin insulators. *Chem Phys Lett* 2005;**417**:22–7.
36. Papageorgiou N, Oughaddou H, Mayne AJ, Dujardin G, Ferro Y, Giovanelli L, et al. Self-assembled molecular chains formed by selective adsorption of Pb-phthalocyanine on InSb (100)-4×2-c(8×2). *Appl Phys Lett* 2003;**82**:2518–20.
37. Bernard R, Comtet G, Dujardin G, Huc V, Mayne AJ. Imaging and spectroscopy of individual CdSe nanocrystals on atomically resolved surfaces. *Appl Phys Lett* 2005;**87**:053114.
38. Bernard R, Comtet G, Dujardin G, Huc V, Mayne AJ, Tang H. Imaging and orientation of individual CdSe nanorods on atomically resolved surfaces. *Phys Rev B* 2007;**75**:045420.
39. Amy F, Enriquez H, Soukiassian P, Brylinski C, Mayne A, Dujardin G. 6H–SiC(0001): an unexpected cubic 4×3 Si phase overlayer. *Appl Phys Lett* 2001;**79**:767–9.
40. Ahn JR, Lee SS, Kim ND, Hwang CG, Min JH, Chung JW. Structural and electronic properties of the Si-rich 6H-SiC(0001) surface. *Surf Sci* 2002;**516**:L529–L534.
41. Baffou G, Mayne AJ, Comtet G, Dujardin G. State selective electron transport through electronic surface states of 6*H*-SiC(0001)-3×3. *Phys Rev B* 2008;**77**:165320.
42. Amy F, Enriquez H, Soukiassian P, Storino P-F, Chabal YJ, Mayne AJ, et al. Atomic scale oxidation of a complex system: O_2/alpha-SiC(0001)-(3×3). *Phys Rev Lett* 2001;**86**:4342–5.
43. Kubo O, Kobayashi T, Yamaoka N, Itou S, Nishida A, Katayama M, et al. Surface reactions of 6H-SiC(0001) 3×3 with oxygen molecules at various temperatures. *Surf Sci* 2003;**529**:107–13.
44. Takami J, Naitoh M, Yokoh I, Nishigaki S, Toyama N. STM and LEED observation of hydrogen adsorption on the 6H-SiC(0001) 3×3 surface. *Surf Sci* 2001;**482–485**:359–64.
45. Preuss M, Bechstedt F, Schmidt WG, Sochos J, Schröter B, Richter W. Clean and pyrrole-functionalized Si- and C-terminated SiC surfaces: first-principles calculations of geometry and energetics compared with LEED and XPS. *Phys Rev B* 2006;**74**:235406.
46. Baffou G, Mayne AJ, Comtet G, Dujardin G. Anchoring phthalocyanine molecules on the 6*H*-SiC(0001) 3×3 surface. *Appl Phys Lett* 2007;**91**:073101.
47. Baffou G, Mayne AJ, Comtet G, Dujardin G, Stauffer L, Sonnet Ph. SiC(0001) 3×3 heterochirality revealed by single-molecule STM imaging. *J Am Chem Soc* 2009;**131**:3210–5.
48. Lorente N. Theory of tunneling currents and induced dynamics at surfaces. In: Hasselbrink E, Lundqvist BI, editors. *Handbook of surface science*, vol. 3. North Holland: Elsevier; 2008. p. 575–620.
49. Wang ZZ, Girard JC, Pasquier C, Jérome D, Bechgaard K. Scanning tunneling microscopy in TTF-TCNQ: phase and amplitude modulated charge density waves. *Phys Rev B* 2003;**67**:121401.
50. Otte AF, Ternes M, von Bergmann K, Loth S, Brune H, Lutz CP, et al. The role of magnetic anisotropy in the Kondo effect. *Nat Phys* 2008;**4**:847.
51. Dobrin S, Harikumar KR, Jones RV, Li N, McNab IR, Polanyi JC, et al. Self-assembled molecular corrals on a semiconductor surface. *Surf Sci* 2006;**600**:L43.
52. Harikumar KR, Polanyi JC, Sloan PA, Ayissi S, Hofer WA. Electronic switching of single silicon atoms by molecular field effects. *J Am Chem Soc* 2006;**128**:16791.
53. Sloan PA, Palmer RE. Two-electron dissociation of single molecules by atomic manipulation at room temperature. *Nature* 2005;**434**:367.

54. Sloan PA, Hedouin MFG, Palmer RE, Persson M. Mechanisms of molecular manipulation with the scanning tunneling microscope at room temperature: chlorobenzene on Si(111)–(7 × 7). *Phys Rev Lett* 2003;**91**:118301.
55. Giessibl FJ. Atomic resolution on Si(111)–(7 × 7) by non contact atomic force microscopy with a force sensor based on a quartz tuning fork. *Appl Phys Lett* 2000;**76**:1470.
56. Sugimoto Y, Abe M, Hirayama S, Oyabu N, Custance O, Morita S. Atom inlays performed at room temperature using atomic force microscopy. *Nat Mater* 2005;**4**:156.
57. Lantz MA, Hug HJ, Hoffmann R, van Schendel PJA, Kappenberger P, Martin S, Baratoff A, Güntherodt HJ. Quantitative Measurement of short-range chemical bonding forces. *Science* 2001;**291**:2580.
58. Sugimoto Y, Pou P, Abe M, Jelinek P, Perez R, Morita S, Custance O. Chemical identification of individual surface atoms by atomic force microscopy. *Nature* 2007;**446**:64.
59. Ternes M, Lutz CP, Hirjibehedin CF, Giessibl FJ, Heinrich AJ. The force needed to move an atom on a surface. *Science* 2008;**319**:1066.
60. Sugimoto Y, Pou P, Custance O, Jelinek P, Abe M, Perez R, et al. Complex patterning by vertical interchange atom manipulation using atomic force microscopy. *Science* 2008;**322**:413–7.
61. Molinàs-Mata P, Mayne AJ, Dujardin G. Manipulation and dynamics at the atomic scale: a dual use of the scanning tunneling microscopy. *Phys Rev Lett* 1998;**80**:3101.
62. Brihuega I, Custance O, Gomez-Rodriguez JM. Surface diffusion of single vacancies on Ge(111)–c(2 × 8) studied by variable temperature scanning tunneling microscopy. *Phys Rev B* 2004;**70**:165410.
63. Mayne AJ, Rose F, Bolis C, Dujardin G. An STM study of the diffusion of a single or a pair of atomic vacancies. *Surf Sci* 2001;**486**:226.
64. Dujardin G, Mayne AJ, Rose F. Surface molecular chain reaction initiated at STM-made individual active sites. *Phys Rev Lett* 1999;**82**:3448.
65. Mayne AJ, Rose F, Dujardin G. An STM study of the growth behaviour of the oxidation of the Ge(111) surface. *Surf Sci* 2003;**523**:157.
66. Gross L, Mohn F, Moll N, Liljeroth P, Meyer G. The chemical structure of a molecule resolved by atomic force microscopy. *Science* 2009;**325**:1110–4.
67. Gross L, Mohn F, Moll N, Meyer G, Ebel R, Abdel-Mageed WM, et al. Organic structure determination using atomic resolution scanning probe microscopy. *Nat Chem* 2010;**2**:1110–4.
68. Gross L, Mohn F, Liljeroth P, Repp J, Giessibl F, Meyer G. Measuring the charge state of an adatom with noncontact atomic force microscopy. *Science* 2009;**324**:1428–31.
69. Foster AS, Barth C, Shluger AL, Reichling M. Unambiguous interpretation of atomically resolved force microscopy images of an insulator. *Phys Rev Lett* 2001;**86**:2373–6.
70. Barth C, Henry CR. Atomic resolution imaging of the (001) surface of UHV cleaved MgO by dynamic scanning force microscopy. *Phys Rev Lett* 2003;**91**:196102.
71. Nimmrich M, Kittelmann M, Rahe P, Mayne AJ, Dujardin G, von Schmidsfeld A, et al. Atomic resolution imaging of clean and hydrogen-terminated C(100)-2 × 1 diamond surfaces using non-contact AFM. *Phys Rev B* 2010;**81**:201403.
72. Tranvouez E, Orieux A, Boer-Duchemin E, Comtet G, Dujardin G, Devillers C, et al. Manipulation of cadmium selenide nanorods with an atomic force microscope. *Nanotechnology* 2009;**20**:165304.
73. Song YJ, Otte AF, Shvarts V, Zhao ZY, Kuk Y, Blankenship SR, et al. Invited review article: a 10 mK scanning probe microscopy facility. *Rev Sci Instrum* 2010;**81**:121101.
74. Levy N, Burke SA, Meaker KL, Panlasigui M, Zettl A, Guinea F, et al. Strain-induced pseudo-magnetic field greater than 300 Tesla in graphene nanobubbles. *Science* 2010;**329**:544–7.

75. Hoffman JE, Hudson EW, Lang KM, Madhavan V, Eisaki H, Uchida S, et al. A Four Unit Cell Periodic Pattern of Quasi-Particle States Surrounding Vortex Cores in $Bi_2Sr_2CaCu_2O_{8+\delta}$. *Science* 2002;**295**:466.
76. Meier F, Zhou L, Wiebe J, Wiesendanger R. Revealing magnetic interactions from single-atom magnetization curves. *Science* 2008;**320**:82.
77. Iacovita C, Rastei MV, Heinrich BW, Brumme T, Kortus J, Limot L, et al. Visualizing the spin of individual cobalt-phthalocyanine molecules. *Phys Rev Lett* 2008;**101**:116602.
78. Baseer Hader M, Pitters JL, DiLabio GA, Livadaru L, Mutus JY, Wolkow RA. Controlled coupling and occupation of silicon atomic quantum dots at room temperature. *Phys Rev Lett* 2009;**102**:046805.
79. Bellec A, Riedel D, Dujardin G, Boudrioua O, Chaput L, Stauffer L, et al. Electronic properties of the n-doped hydrogenated silicon (100) surface and dehydrogenated structures at 5K. *Phys Rev B* 2009;**80**:245434.
80. Bellec A, Riedel D, Dujardin G, Boudrioua O, Chaput L, Stauffer L, et al. Non-local activation of a bistable atom through a surface state charge-transfer process on Si(100)-(2 × 1). *Phys Rev Lett* 2010;**105**:048302.
81. Kitchen D, Richardella A, Tang JM, Flatte ME, Yazdani A. Atom-by-atom substitution of Mn in GaAs and visualization of their hole-mediated interactions. *Nature* 2006;**442**:436.
82. Wintterlin J, Bocquet M-L. Graphene on metal surfaces. *Surf Sci* 2009;**603**:1841–52.
83. Brar V, Decker R, Solowan H-M, Wang Y, Maserati L, Chan KT, et al. Gate-controlled ionization and screening of cobalt adatoms on a graphene surface. *Nat Phys* 2011;**7**:43.
84. Yang H, Mayne AJ, Boucherit M, Comtet G, Dujardin G, Kuk Y. Quantum interference channelling at graphene edges. *Nano Lett* 2010;**10**:943.
85. De Padova P, Quaresima C, Ottaviani C, Sheverdyaeva PM, Moras P, Carbone C, et al. Evidence of graphene-like electronic signature in silicene nanowires. *Appl Phys Lett* 2010;**96**:261905.
86. Lalmi B, Oughaddou H, Enriquez H, Kara A, Vizzini S, Ealet B, et al. Epitaxial growth of a silicene sheet. *Appl Phys Lett* 2010;**97**:223109.
87. Maier M. Ultimate nanoprobing. *Pico (Omicron NanoTechnol Newsl)* 2010;**14**:2–4.
88. Wolkow RA. Controlled molecular adsorption on silicon: laying a foundation for molecular devices. *Annu Rev Phys Chem* 1999;**50**:413–41.
89. Joachim C, Gimzewski JK, Aviram A. Electronics using hybrid-molecular and mono-molecular devices. *Nature* 2000;**408**:541–8.
90. Joachim C, Ratner MA. Molecular electronics: some views on transport junctions and beyond. *Proc Natl Acad Sci USA* 2005;**102**:8801–5.
91. Guchet X. Nature and artifact in nanotechnologies. *Int J Philos Chem* 2009;**15**:5–14.

Part I

Electronics at the Atomic-Scale

Chapter 2

STM Manipulation of Single Atoms and Molecules on Insulating Films

Jascha Repp*
*Institute of Experimental and Applied Physics, Universität Regensburg, Germany

Gerhard Meyer†
†IBM Research-Zurich, Rüschlikon, Germany

1. INTRODUCTION

Atomic and molecular manipulation is one of the most fascinating applications of the scanning tunnelling microscope (STM). Experimental realizations are the construction of atomically precise nanostructures[1] and single-molecule chemistry,[2] for example. A large variety of manipulation techniques have been demonstrated based on either direct force interactions between the adsorbate and tip, the electric field in the tunnelling gap, or inelastic tunnelling. It is well recognized that the adsorbate–substrate interaction plays a predominant role in the yield and controllability of the different techniques. For instance, on low-indexed metal surfaces, lateral manipulation of single atoms or molecules is very efficient to build up complex nanostructures, whereby the adsorbate is pushed or pulled over the surface by the forces exerted by the STM tip. In contrast, on semiconductors, processes based on inelastic tunnelling are preferably used because of the more directional nature of the bonding.[3] As STM relies on a finite conductance of the tip–sample junction, only very few STM studies dealt with insulating surfaces until recently. In recent years, however, ultrathin insulating films of wide-band-gap insulators have been used in STM experiments, with the aim of decoupling adsorbate electronic states from a conductive substrate. As many physical and chemical properties of and interactions between adsorbates on insulating surfaces differ qualitatively from those on conducting surfaces, these studies uncovered many novel phenomena. Bulk insulators have no propagating electronic states in the band gap region and thus are not suited for STM, whereas ultrathin insulating films, only a few atomic layers thick, allow the tunnelling of electrons at measurable rates.

Frontiers of Nanoscience, Vol. 2. DOI: 10.1016/B978-0-08-096355-6.00002-7

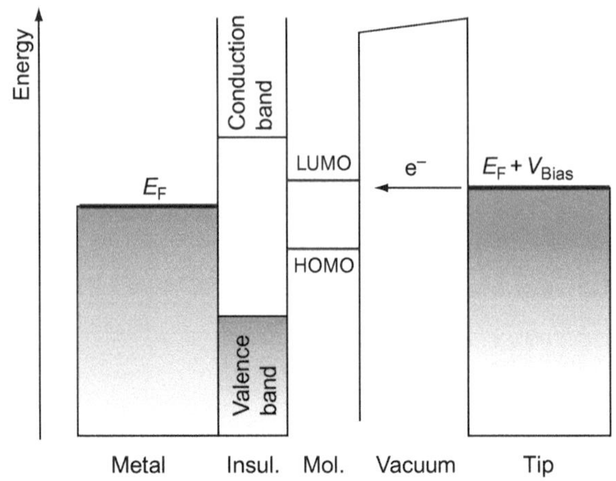

FIGURE 2.1 Schematic view of a double-barrier tunnelling junction geometry. Because of the separation by an ultrathin insulating film, the electronic coupling of adsorbates to an underlying conducting substrate is only weak.

The atoms/molecules are electronically only weakly coupled to the underlying metallic substrate so that the STM junction represents a double-barrier tunnelling junction (Figure 2.1). The novel phenomena that could be observed include tunnelling-induced fluorescence,[4,5] vibronic spectroscopy,[6] (time-resolved) spin-flip spectroscopy on single metal atoms and engineered atomic nanostructures,[7–9] manipulation of the charge state of metal atoms[10] and molecules,[11] orbital imaging and engineering,[12,13] bipolar tunnelling,[14] and molecular switching based on hydrogen tautomerization.[15] In the following, we will review recent STM-based atomic and molecular manipulation experiments on insulating films. The review will start with a short overview of suitable insulating film materials and their growth, followed by a detailed discussion of specific manipulation processes and related work on single atoms and molecules.

2. ULTRATHIN INSULATING FILMS

A prerequisite for STM studies on adsorbates on insulating films is the ability to grow stable and atomically thin insulating films on metal or semiconductor surfaces with a well characterized and ordered geometric structure. Whereas later we will focus on alkali halide films, we will first present a brief overview of various corresponding systems that have been studied. Oxide films certainly belong to the most intensively studied systems, in particular, because of their great importance as substrates in the investigation of catalytic processes.[16,17] The most prominent example is probably the growth of a Al_2O_3 film on a NiAl (110) alloy substrate[18] by oxidation of the NiAl(110) surface. The high bulk melting temperature of NiAl, as compared to that of a pure aluminium substrate, enables the use of higher annealing temperatures, which results in a well-ordered one-monolayer-thick film. Recently, also a detailed understanding of the large and complicated unit cell structure was gained.[19] Other examples of studies with monolayer oxide films are the oxidation of Cu(110)[20] and the formation of flat PbO films on Pb(111).[21] By evaporation of a metal in the presence of gaseous oxygen, films with variable thicknesses can be prepared, as was demonstrated for MgO.[22] Apart from oxide films, also nitrides (CuN/Cu(100); Ref. 8) and strongly polar fluorides have been studied (CaF_2/Si(100); Refs. 23–25). In contrast to oxides, alkali halides can be directly deposited by thermal evaporation. The high binding energy of ion pairs results in the evaporation of the material as multiples of alkali halide pairs, which retains the correct stoichiometry of the film. Several STM and atomic force microscopy (AFM) investigations using NaCl on Al(111) and Al(100),[26] NaCl on Cu(111), Cu(100), Cu(311), Cu(211), Ag(100)[4,27–31] and NaBr/NiAl(110)[32] showed that two-dimensional (100)-oriented NaCl islands with a thickness of only a few atomic layers can be reliably grown on metal surfaces. On semiconductor surfaces, epitaxial growth of NaCl was demonstrated on Ge,[33–35] InSb[36] and GaAs substrates.[37] All these materials are strongly polar, but electronic decoupling can, in principle, also be achieved using nonpolar materials as was demonstrated, for example, using thin layers of organic molecules[38,39] or even of noble gas.[40]

In the following, we will discuss the growth of NaCl films on flat low-index Cu(111), Cu(100) substrates and a regularly stepped Cu(311) surface at varying deposition temperatures in more detail. In all three cases, (100)-oriented two-dimensional NaCl islands having a thickness of only a few atomic layers could be grown.

The growth behaviours observed for the Cu(111) and the Cu(100) substrate orientation are very similar: for deposition temperatures of room temperature and higher, the NaCl islands start with a double-layer thickness and perfect nonpolar step edges, in which the anions and the cations alternate (cf. Figure 2.2). The diameter of the islands ranges from a few hundred nanometres to several micrometres. On top of the initial double layer, smaller islands of additional layers are formed. Substrate defect steps are smoothly overgrown in the so-called carpet-like growth.[35] For a deposition temperature of about 200 K, double- and single-layer islands were found, with both island types being smaller than those found for a deposition temperature of $T \geq 300$ K. On both the Cu(111) and the Cu(100) substrate surface, the growth is incommensurate and many different orientations of the NaCl(100) islands with respect to the copper crystal could be observed. As a consequence, different Moiré patterns can be observed in atomically resolved STM images of mono- and bilayers of NaCl on these substrates (cf. Figure 2.3). In both cases, there is a preference for orientations in which one of the polar NaCl $\langle 01\bar{1} \rangle$ directions is almost parallel to one of the close-packed $\langle 01\bar{1} \rangle$ directions of the copper substrates. In the case of NaCl(100) on Cu(100), a perfect alignment of these directions would support a commensurate structure because of the close 2:3 match of the Cu and NaCl lattice constants. This would result

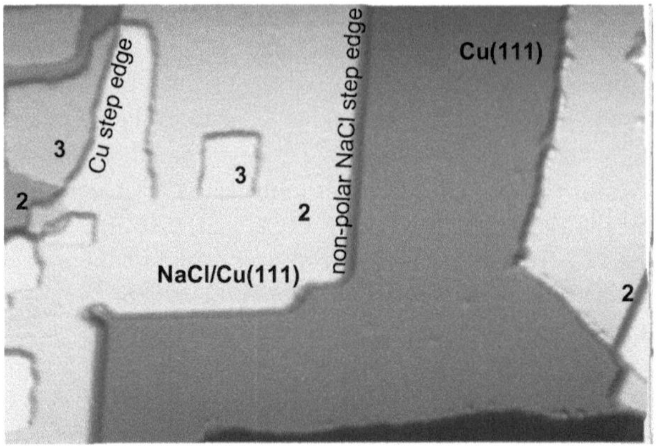

FIGURE 2.2 NaCl island on Cu(111). The numbers indicate the NaCl film thicknesses in atomic layers. The island starts with a double layer and has straight nonpolar step edges. Substrate defect steps are smoothly overgrown. (210×250 nm; 588 mV; 0.17 nA)

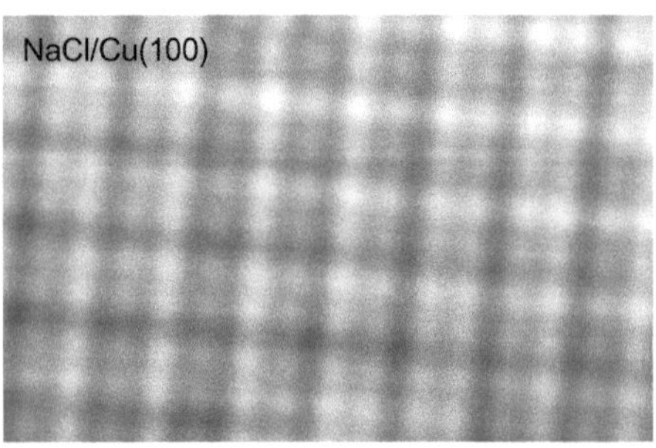

FIGURE 2.3 Atomically resolved double-layer island of NaCl on Cu(100). For this specific island, the polar direction of the NaCl film is well aligned with one of the close-packed Cu(100) directions. Although in this orientation, the good 2:3 lattice match would support a (3×3) superstructure, the growth is incommensurate, and the atomic corrugation exhibits a Moiré pattern (60×40 Å; 24 mV; 0.22 nA)

in a (3×3) superstructure with respect to the Cu(100) substrate surface. Despite this good lattice match, however, the growth is observed to be incommensurate and the orientational alignment is not perfect, as can be deduced from the Moiré pattern shown in Figure 2.3. However, the mere possibility of a (3×3)-superstructure formation makes density-functional theory (DFT) simulations in a supercell geometry of this system possible.

On the intrinsically stepped Cu(311) substrate surface, the growth behaviour differs[30] from that observed on the flat substrate surfaces. For deposition temperatures above 400 K, flat, (100)-oriented NaCl islands of monoatomic height could be observed. For coverages of $0 < \Theta < 2$ monolayers, we find that the second-layer growth of NaCl starts only after the first layer has been completed. For higher coverages, the growth slightly deviates from perfect layer-by-layer growth, and the third and fourth layer grow before the second layer has been completed. This indicates a strong substrate–overlayer interaction. STM measurements further indicate that NaCl grows pseudomorphically, with one of its polar $\langle 01\bar{1} \rangle$ directions parallel to the intrinsic Cu steps.

Perpendicular to the intrinsic steps, the lattice constant of the NaCl adlayer is adapted to that of the substrate, which leads to a stretching of more than 6% with respect to the bulk value. In this case, the Cl^- ions are located above the intrinsic steps of the Cu(311) surface, whereas the Na^+ ions are located between the steps. This contrasts the growth behaviour on Cu(100), on which a much smaller misfit did not result in a commensurate structure. This growth behaviour was explained using a model based on an inherent attribute of regularly stepped metal surfaces, namely the charge modulation due to the Smoluchowski effect.[30] In his pioneering work, Smoluchowski[41] discussed two effects responsible for the charge distribution at metal surfaces. The first effect is the relatively slow decrease of the negative charge density perpendicular to the surface. The second effect, the so-called smoothing, implies that conduction electrons do not completely follow the short-range corrugation of the surface. Smoluchowski discussed how the surface corrugation affects the work function, but it also leads to a lateral charge modulation of the substrate surface. At the highly corrugated Cu(311) surface, one expects a positive charge along the intrinsic steps and a negative charge between the steps. The reason for the exceptional stability of a NaCl/Cu(311) interface becomes immediately apparent (cf. Figure 2.4): columns of Cl^- and Na^+ ions align with stripes of opposite charge located at steps and troughs, respectively, resulting in a strong electrostatic interaction between the NaCl adlayer and the Cu(311) substrate. Apparently, the interaction is strong enough to stabilize a stretching of more than 6% perpendicularly to the Cu rows. From our model[30], we concluded that, for a regularly stepped surface with a suitable geometry, an energetically favourable interface matching will be achieved if the polar columns of the ionic layer fit with the intrinsic steps of the metal substrate. Later, DFT calculations corroborated this conclusion and also showed that the bonding of the overlayer on the surface is further stabilized by the formation of weak chemical bonds between the step Cu atoms and the Cl^- ions.[42] Independently of the relative contribution of chemical or electrostatic interactions, these findings emphasize the important role of vicinal

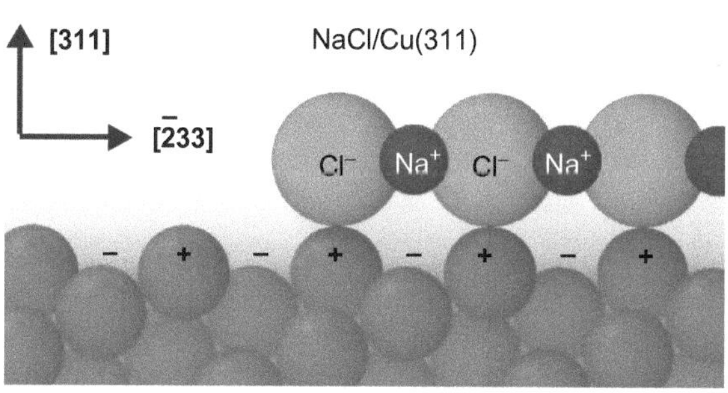

FIGURE 2.4 Basic model of the binding geometry for NaCl/Cu(311): the Smoluchowski smoothing effect leads to a charge-corrugated surface, forming a strong bond with NaCl.

metal surfaces play in the growth of ionic thin films. Moreover, this effect could also be exploited to find ways to obtain atomically flat films of other polar insulators that tend to grow in three-dimensional islands on flat metal surfaces.

3. INTERFACE STATE IN NaCl/Cu(111)

The well-known Cu(111) surface-state band survives NaCl adsorption and forms an interface state (IS) band that is confined to the insulator/metal interface.[31] In STM images, the IS manifests itself as the typical standing-wave patterns[43,44] arising from the scattering of the electrons (cf. Figure 2.5). The dispersion relation of this IS differs from that of the clean Cu(111) surface, as manifested by a larger Fermi wavelength of $\lambda_F = 3.8$ nm (clean Cu: $\lambda_F = 3.0$ nm). The dispersion was determined from differential conductance (dI/dV) images taken at various bias voltages, as was done for the clean Cu surface.[43–45] The band minimum was obtained by recording the dI/dV signal at a fixed position as a function of energy,[45,46] and it is shifted upward in energy by 230 ± 30 meV for the adsorption of an insulating NaCl overlayer.

The resulting parabolic dispersion $E = E_0 + (\bar{h}k)^2/2m^*$ is slightly wider, that is, the effective mass m^* has increased from $(0.40 \pm 0.02)m_e$ to $(0.46 \pm 0.04)m_e$, where m_e denotes the free-electron mass. Of interest in the study of adsorbates on NaCl films is the observation that the IS can be used as a probe to determine the charge state of any defect in the NaCl film or of an adsorbate on it.[47] To infer the charge state from observations related to the IS, it is important to have qualitative knowledge of the probability distribution of the IS electrons perpendicular to the surface. This probability distribution can be calculated in using the one-dimensional phase-accumulation model.[48,49] In this analysis, the wave function Ψ of surface electrons is found by matching the wave-function phase of the analytical solution inside the Cu crystal[50] to

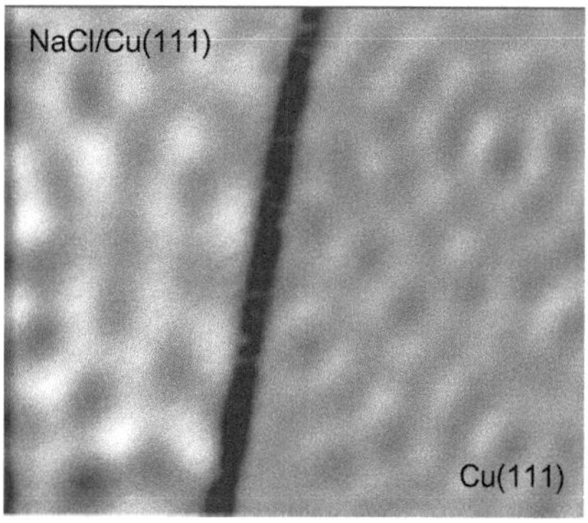

FIGURE 2.5 Differential conductance (d*I*/d*V*) STM image at low bias voltage (+20 mV), showing standing-wave patterns of the interface state electrons in a bilayer of NaCl on Cu(111) (left panel). The wavelength is slightly larger than that for the clean Cu (111) surface area (right panel).

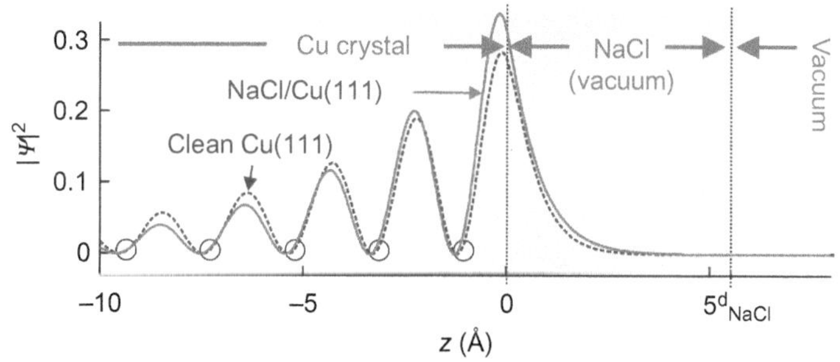

FIGURE 2.6 Probability distribution of the interface state electrons perpendicular to the surface for a bilayer of NaCl on Cu(111) and the clean Cu(111) surface calculated using the phase-accumulation model. Adapted from Ref. 31.

that of the numerical solution for the outside region, obtained by integrating the Schrödinger equation.[49] For the latter, an electrostatic potential had to be assumed: for the clean Cu surface, this is given by the image potential, whereas for NaCl/Cu(111), the modified image potential[48] as well as the lowering of the work function upon NaCl adsorption was considered.[51] The image potential can be treated classically, as done in Ref. 48, by applying the static dielectric constant of NaCl. The nonphysical divergence of the image potential was suppressed. The average barrier potential in the NaCl layer is given by the upper edge of the band gap. This choice of electrostatic potential is somewhat arbitrary, but the main results of this analysis do not depend critically on the details of the potential chosen.

The resulting probability distribution is plotted in Figure 2.6: it is barely modified upon adsorption of NaCl. Thus, the main contribution of the probability distribution is located within the topmost few layers of the Cu crystal. The probability distribution falls off exponentially within the NaCl adlayer, such that it is negligibly small on top of a NaCl double layer. Consequently, the short-range potential of a neutral adsorbate should be a weak scatterer of the IS electrons. A charged adsorbate, in contrast, results in a long-range electrostatic potential that should strongly scatter the IS electrons below it. A positively charged adsorbate will, in addition, create a local attractive potential for the IS electrons underneath. In general, whenever a two-dimensional, free-electron-like interface band encounters an attractive potential, it will form a bound state, that is, an IS localization.[47,52,53] Thus, the charge state of any defect or adsorbate can unambiguously be determined to be neutral, negative, or positive, depending on whether scattering of the IS and an IS localization can be observed.

4. MANIPULATION OF METAL ATOMS

4.1. Controlling the Charge State

In Section 1, we stated that many physical and chemical properties of adsorbates on an insulating surface differ qualitatively from those of adsorbates on a conducting surface. This will be substantiated in the discussion of the charge state control of individual gold adatoms on NaCl/Cu(111)[10] below. By positioning the STM tip above an Au adatom and applying a voltage pulse, the adatom can be reversibly switched between its neutral and its negatively charged state. Most importantly, both states are stable: an additional charge will remain on the adsorbate until it is removed by a voltage pulse of reversed sign.

Upon adsorption, individual Au adatoms on NaCl(100)/Cu(111) are imaged as protrusions. The apparent height is in the range of 2.0–2.5 Å for a tunnelling current of $I = 10$ pA to 0.2 nA. Note that this apparent height is rather large compared with that of individual metal atoms on metal substrates. The adsorption site was determined to be on top of the Cl^- ions, both directly from atomically resolved STM images of the Cl^- ions and indirectly from the adatom position with respect to artificially created Cl vacancies.

As electrons can still tunnel through the ultrathin NaCl film, the charge state of the gold atoms is not *a priori* obvious.[54] The absence of scattering of the IS electrons at the Au adatoms indicates, however, that the Au adatoms remain neutral upon adsorption, and we will thus refer to them henceforth as Au^0 (cf. Section 3 and Figure 2.8).

By applying a voltage pulse, the adsorption state of an individual Au adatom can be manipulated. In this adatom manipulation, shown in Figure 2.7, the tip is first positioned directly above an Au adatom. The feedback loop is then switched off, and a positive voltage $V \geq 0.6$ V is applied. After a certain time t, which depends on the specific tunnelling parameters, a sharp current drop by about a factor of 3 is observed. In subsequent images, the corresponding Au adatom appears differently, but is still located at the same position. The manipulated Au adatom is imaged as a sombrero-like shape, with an about 0.5 Å smaller protrusion than the original state surrounded by a depression. By applying a similar, but negative voltage pulse of about -1 V, the manipulated adatom can be switched back to its original state. As can be seen in Figure 2.8, the gold adatom in the manipulated state scatters

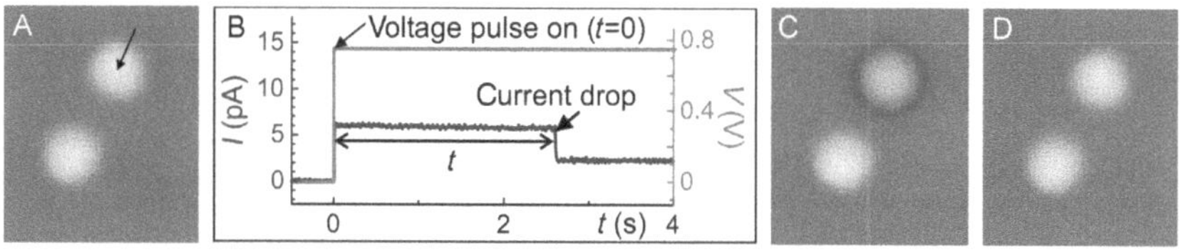

FIGURE 2.7 Manipulation of the Au adatom state. After recording image (A), the STM tip was positioned above one of the Au adatoms (arrow), and a positive voltage pulse was applied to the sample. After a certain time *t*, a sharp decrease of the tunnelling current can be observed (B). A subsequent STM image (C) shows that the manipulated Au adatom did not change its position but has a different appearance. By applying a negative voltage pulse, the manipulated adatom can be switched back to its initial state (D). Adapted from Ref. 10 (50 mV; 10 pA).

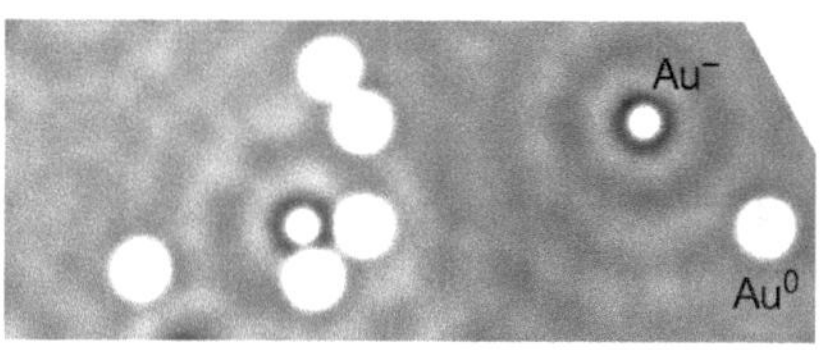

FIGURE 2.8 High-contrast STM image of six neutral and two negatively charged Au adatoms on a bilayer of NaCl on Cu(111). The standing-wave patterns show that the charged Au adatoms scatter the interface state electrons, whereas the neutral adatoms do not (76 mV, 0.2 pA).

the IS electrons at the NaCl/Cu(111) interface, but no IS localization at the gold adatoms can be observed. This indicates that the manipulated gold atoms are negatively charged (Au^-).

The states of the neutral and the negative Au adatoms are both stable and must therefore be associated with two different geometric configurations of the adatom and the NaCl film. A simple electron transfer without lasting changes of the ion core positions would not be stable because the electron residing in an excited state on the manipulated Au adatom would rapidly tunnel into the metal.

DFT simulations[31] provided deeper insight into the underlying principle of the experimentally observed charge bistability: in agreement with our experiments, the theoretical investigation found two different stable states for Au atoms in on-top sites of a Cl^- ion in the NaCl film on Cu(100), namely a neutral and a negatively charged state. Whereas the gold atom in its neutral state leaves the ionic positions within the NaCl film relatively unperturbed, the negatively charged gold adatom induces large ionic relaxations within the NaCl film. The Cl^- ion underneath the adatom is forced to move downwards by 0.6 Å and the surrounding Na^+ ions upward by 0.6 Å. This relaxation pattern creates an attractive potential for the additional charge on the Au adatom, which is further stabilized by the screening charge in the metal substrate and by the electronic polarization of the ionic layer. This shifts the Au (6s) state, which is singly occupied in a neutral gold adatom, downwards by 1.0 eV, such that it is below the Fermi level and becomes doubly occupied in the negatively charge gold adatom.

To find out more about the physical mechanism behind the neutral-to-negative charge state manipulation process, the quantum yield was determined as a function of the voltage applied during the manipulation. In STM manipulation, this quantum yield is simply defined as the probability of a manipulation event per tunnelling electron. As will be discussed in detail below, the quantum yield can reach values of order unity for voltage pulses of $V \geq 1.4$ V. This extremely high yield is consistent with a capture of electrons tunnelling resonantly into the negative ion-resonance (NIR) state. (The NIR is the negatively charged adsorbate with the system in the same ionic configuration as for the neutral adsorbate.) This suggests that, because of the presence of the insulator, the lifetime of the NIR state of the adatom is in the range of the ionic vibrational periods, which results in a capture of the tunnelling electrons. An electron tunnelling resonantly into the NIR state remains there for a sufficiently long time so that the adatom and its surrounding ions have time to relax and the NIR state to shift below E_F, and the electron is captured. This model implies that the NIR is 1.4 eV above the Fermi level or 2.6 eV below the vacuum level,[51] which is close to the electron affinity of an isolated Au atom of 2.3 eV (below the vacuum level).

For voltage pulses V smaller than 0.9 V, the quantum yield was determined directly by measuring the current pulse and was found to increase exponentially with increasing voltage (cf. Figure 2.9A). However, the ratio of current noise to bandwidth of the current amplifier in a low-temperature STM set-up limits the direct determination of the quantum yield to values of below 10^{-5}. Therefore, to roughly determine the yield also for $V > 0.9$ V, for which it reaches higher values, an alternative technique to quantify the

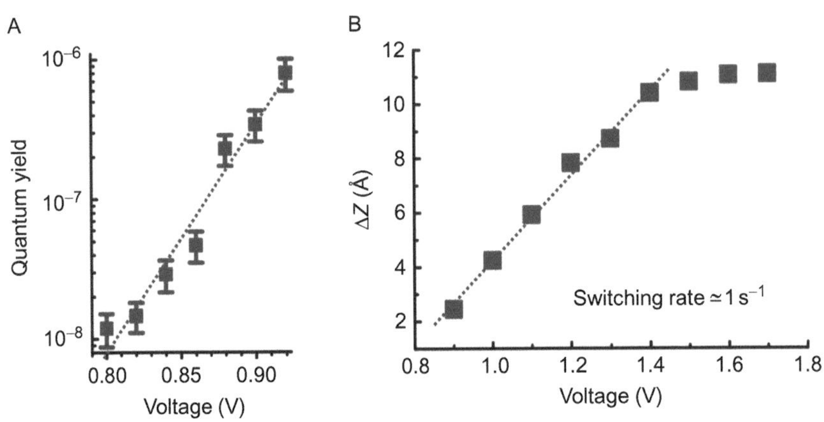

FIGURE 2.9 Statistical analysis of the switching behaviour of Au adatoms on an NaCl bilayer on Cu(111). (A) For voltage pulses below 0.9 V, the quantum yield is sufficiently low so that it can be determined directly by measuring the current. In this voltage range, the yield increases exponentially with voltage (line fit). (B) For voltages in the range of 0.9–1.7 V, the tip was retracted by Δz from its initial position z_0, corresponding to $I_0 = 10$ pA at $V_0 = 500$ mV, such that a mean switching time of 1 s could be observed. Saturation is observed for higher voltages, whereas for voltages below 1.4 V, Δz increases linearly with the voltage (line fit). Adapted from Ref. 10.

manipulation behaviour had to be applied. We determined the distance Δz, the tip has to be retracted from an initial position z_0, corresponding to $I_0 = 10$ pA at $V_0 = 0.5$ V, so that a mean switching time of 1 s is observed. For this analysis, the tip was positioned above an Au adatom using $I_0 = 10$ pA at $V_0 = 0.5$ V as scanning parameters. Then the feedback loop was switched off, and the tip retracted from the surface by the distance Δz. After a short pause to minimize creep of the piezo elements, a voltage pulse of 0.1-s duration was applied. As described above, the current flow during the pulse was too small to be measured directly. Therefore, after the voltage pulse has been applied, the tip was moved back to its initial scanning position, where at $V_0 = 0.5$ V, a sizeable current revealed whether the charge state had changed. This procedure was repeated if necessary, and Δz was chosen in such a way that on average ten pulses of 0.1-s duration were needed for a successful manipulation.

The result of this statistical analysis (Figure 2.9B) reveals that below 1.4 V, Δz increases linearly with voltage, whereas for higher voltages, saturation is observed.

The linear increase of Δz below 1.4 V is consistent with an exponential increase of the yield. The saturation rules out that the electric field in the tunnelling junction is responsible for the switching. To estimate the quantum yield at saturation, that is, for $\Delta z = 10.5$ Å and $V = 1.4$ V, the tunnelling current I needs to be extrapolated from $I_0 = 10$ pA at $\Delta z = 0$ Å and $V_0 = 0.5$ V. Using an exponential extrapolation of I with Δz yields a decrease of about 10 orders of magnitude, which was quantified using an I (z) measurement at higher currents. There should also be a dramatic increase of the tunnelling current with increasing voltage because of tunnelling through the NIR state. Note that I (V) cannot be measured over this voltage range without immediately switching the Au adatom. Based on I (V) measurements for tunnelling through the NIR of molecules[12] and on defects[47] in NaCl films, one can estimate that tunnelling through the NIR state increases the tunnelling current by about two orders of magnitude. The extrapolation of I at $\Delta z = 10.5$ Å and $V = 1.4$ V thus yields a current on the order of one electron per second, corresponding to a quantum yield of order unity.

The difference in physical properties of the two states was documented in diffusion experiments. Interestingly, the two differently charged states of the Au adatom are conserved during the diffusion process. In the Au^0 state, adatom diffusion sets in at a temperature of about 60 K, whereas in the Au^- state, the adatoms already diffuse at a lower temperature of 40 K. This difference even allows the diffusion of single Au adatoms to be switched on and off.

Similar to the Au adatom charge manipulation, that is, $Au^0 \rightarrow Au^-$ on a bilayer NaCl(2ML)/Cu(111) and Cu(100),[10] so can Ag^0 be switched to a negatively charged adatom state, Ag^-, by electron tunnelling resulting from the application of a positive voltage pulse ($V \geq 1.3$ V). As for an Au adatom, the formation of Ag^- results in both a sudden drop in the current signal by about a factor of three and an image that is characterized by a 2.0-Å high protrusion, surrounded by a faint depression (sombrero shape). The switching is also reversible, that is, Ag^- can be transformed back into Ag^0 by hole tunnelling by application of a negative voltage pulse ($V \leq -0.2$ V). Note that because of the Moiré pattern, Ag^- is unstable for some sites and will spontaneously switch back to Ag^0 before the Ag^- can be imaged. On trilayer NaCl (3ML)/Cu(100), this spontaneous neutralization always occurs.

In contrast to an Au adatom, an Ag adatom can also be manipulated into a positively charged adatom state Ag^+ through hole tunnelling by application of a negative voltage pulse ($V \leq -1.3$ V). This manipulation is possible for all adsorption positions of Ag^0 with respect to the Moiré pattern and results in a stable Ag^+. Images of an Ag^+ are characterized by a small, elongated protrusion of only about 0.5 Å in height. Electron tunnelling induced by application of a positive pulse ($V \geq 1.5$ V) will not only neutralize the Ag^+ again but also cause its desorption.

The adsorption sites of Ag^0, Ag^-, and Ag^+ were determined experimentally from the STM images. In atomically resolved images,[26,55,56] Ag^+ was found at a site bridging two Cl anions. Using the Ag^+ as a marker, the adsorption sites of Ag^- and Ag^0 were both found to be on top of the Cl anions. The orientation of the elongated shape of the Ag^+ protrusion in the image will differ for the two bridge sites in a surface unit cell of NaCl(100). From the site determination, it can be deduced that the protrusion is elongated towards the two nearest neighbouring Cl anions.

To provide further justification for the assignment of the charge states of the different Ag adatom states, the same scheme[47] as described above can be employed based on the observed behaviour of the interaction of IS electrons of NaCl/Cu(111) with the Ag adatoms. On this surface, Ag adatoms exhibit a similar switching behaviour as on NaCl/Cu(100), except that here Ag^- is not stable and its interaction with the IS electrons cannot be studied. This difference of the Ag^- stability was attributed to the larger work function of NaCl/Cu(111). The identification of a localized IS in the dI/dV spectrum (Figure 2.10F) from Ag^+ shows that it is positively charged.[52,53] In the case of Ag^0, both the absence of a localized IS and a much weaker standing-wave pattern of IS electrons than from Ag^+, as identified from the high-contrast STM images, are consistent with Ag^0 being neutral (Figure 2.10E and F). Finally, the similarities of the manipulation procedure and STM images of Ag^- with the Au adatom charge manipulation[10] indicate that Ag^- is negatively charged.

4.2. Vertical Transfer and Build-up of Nanostructures

Unfortunately, atomically precise lateral manipulation of single metal atoms, as routinely applied on low-index metal substrates, has so far not been successful on insulating films. Therefore, scientists had to search for other

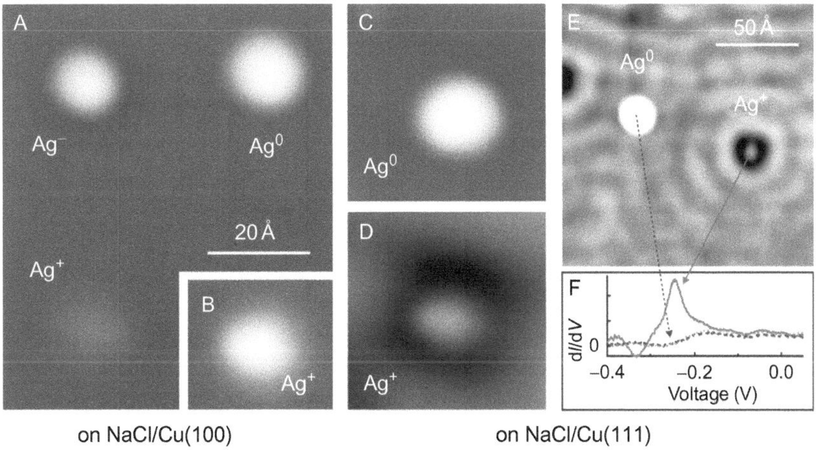

FIGURE 2.10 STM images and spectra of different Ag adatom states (Ag^0, Ag^+, and Ag^-) on NaCl(2ML)/Cu(100) (A and B) and Cu(111) (C–F). In (A) and (C), the maximum scale (white) corresponds to a height of 2.5 Å; (B) and (D) Ag^+ with a fourfold higher contrast. An image at even higher contrast (E) and local d*I*/d*V* spectra (F) reveal scattering of interface state electrons and an interface state localization at Ag^+ but not at Ag^0. The solid, dashed, and dotted spectra refer to Ag^+, Ag^0, and bare NaCl/Cu, respectively. The following bias voltages and tunnelling currents were used: (A) 50 mV, 1 pA; (B) 200 mV, 0.5 pA; (C) −558 mV, 62 pA; (D) −211 mV, 62 pA; and (E) 15 mV, 0.8 pA.

manipulation techniques to build up atomic-scale structures. The group led by Heinrich used copper nitride monolayers on copper as substrate system to study the spin-related properties of individual transition-metal adatoms[7,57] as well as monatomic metal wires consisting of up to 10 Mn adatoms.[8] The copper nitride monolayers were important for decoupling the adsorbate spins from the underlying metal. To construct the monatomic metal wires on these Cu_2N films, they developed a vertical manipulation technique in analogy to the vertical manipulation of Xe,[58] in which the atom to be transferred is excited inelastically and follows the direction of the electrons. To pick up a metal atom, they positioned the tip over the atom with the feedback loop closed at 10 mV and 1 nA. Then they opened the loop and approached the tip by 2.0 Å to the surface while applying a very low bias of 1 mV. After a pause of 1 s, the tip voltage was increased to +2.0 V, which initiated the atom transfer within typically 200 ms. Finally, the tip height and voltage were set to the previous scanning values. Putting an atom down is done in the same way, the only difference being the vertical tip displacement (2.2 Å) and the transfer voltage (−0.5 V). Note that the success rate of this procedure depends strongly on the microscopic geometry of the tip. However, repeated tip pokes have reproducibly generated tips that could reliably pick up and drop off a specific atom tens of times in a row. Although magnetic atoms can be encountered both on Cu and on N sites, they always land on an N site in the drop-off procedure. As a result, the potential drop-off sites are 3.6 Å apart. This turns out to be sufficient to allow the exact landing site to be selected with practically 100% accuracy before putting down an atom, provided that the tip is in a good condition for manipulation. The monatomic Mn wires produced with this technique are shown in Figure 2.11.

5. IMAGING MOLECULAR ORBITALS

A molecule that electronically is only weakly coupled to a metallic substrate will retain its discrete energy level structure and allow its frontier orbitals to be imaged. The ability to image individual orbitals is the basis of molecular orbital engineering, which will be discussed later. In this section, we present the prerequisites of orbital imaging in greater detail. As an example, we will use pentacene, which is one of the best-studied organic semiconductors and exhibits a relatively small band gap and a high mobility. It is also ideally suited for single-molecule experiments with the STM because it is a molecule with a delocalized electronic system that is planar, chemically stable, and can easily be deposited with high purity by thermal evaporation in a ultra-high vacuum environment.[12]

As shown in Figure 2.12, individual pentacene molecules on a two-layer NaCl film on Cu(111) appear as featureless, rod-like protrusions, 1.2 Å in height. An adsorption-site determination using co-adsorbed Au atoms

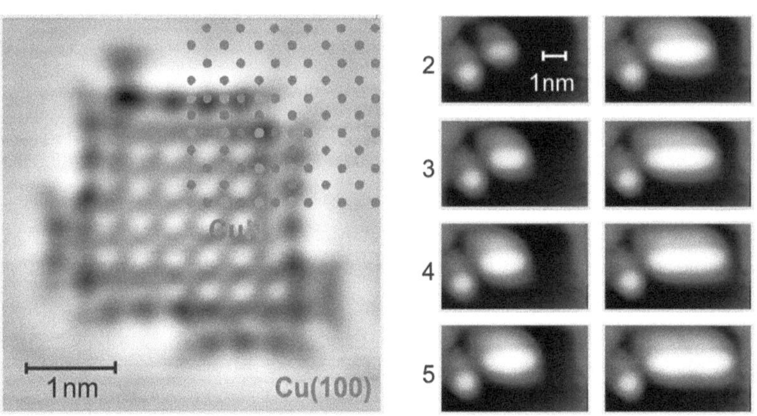

FIGURE 2.11 Mn wires on CuN. Left image: STM image (10 mV, 1 nA) of a CuN island on Cu (100). It has been high-pass-filtered to enhance contrast and the lattice positions of Cu (red dots) and N (blue dots) atoms have been overlaid. Right image: sequence of STM images showing how a wire of Mn atoms is built up from two to nine atoms in length on CuN (10 mV, 0.1 nA). The Mn atom in the lower left of each image appears as a 0.34-nm high bump. The dimer has a smaller peak height of 0.30 nm. Wires with four or more atoms have peak heights of 0.45 nm. Individual atoms in the wire cannot be resolved. Artefacts seen to the upper left of all structures are characteristic of the atomic arrangement of the tip used for manipulation. Reprinted with permission from Ref. 8.

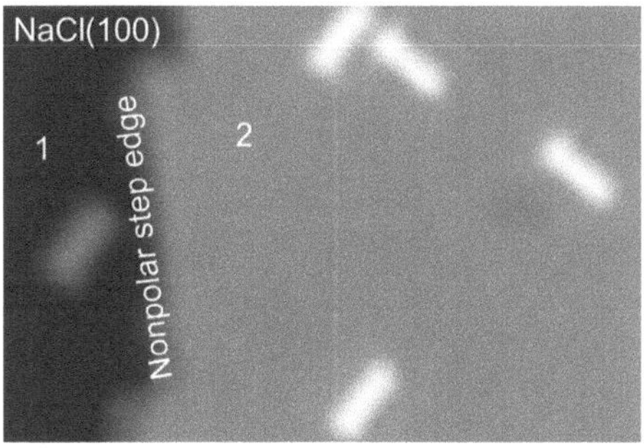

FIGURE 2.12 Individual pentacene molecules on a single-/double-layer NaCl film on Cu(111). The pentacene molecule is located with its centre on top of a Cl ion of the NaCl film, and its long axis is aligned parallel to one of the polar $\langle 01\bar{1} \rangle$ directions of the NaCl(100) films. Adapted from Ref. 12 (500 mV, 11 pA).

revealed that the centre of the pentacene molecule is positioned on top of a Cl− ion of the NaCl layer and that the long axis of the molecule is aligned parallel to one of the polar $\langle 01\bar{1} \rangle$ directions of the NaCl films. Scanning tunnelling spectroscopy (STS) on an individual pentacene molecule exhibits two distinct features in the dI/dV signal, centred around −2.4 and 1.7 V, and a broad gap region of low conductance between these peaks (Figure 2.13). Bias-dependent imaging shows that directly correlated to these spectroscopic features three qualitatively different STM images can be obtained: within the gap region, a pentacene molecule appears as a featureless rod, as described above. In contrast, for voltages exceeding the peak positions, the molecule appears much wider and higher and shows internal structure (panels to the left and right). The apparent height increases to 2.0 and 3.5 Å for voltages below −2.4 and above 1.7 V, respectively. The spatial resolution can be strongly enhanced by picking up a pentacene molecule with the STM tip (lower row of STM images in Figure 2.13), henceforth referred to as pentacene-terminated tip. In this case, the image acquired at voltages in the gap region shows five faint protrusions separated by nodal planes and 2.3 Å apart. The corresponding STM images at voltages below −2.4 and above 1.7 V very closely resemble the native highest occupied (HOMO) and lowest unoccupied molecular orbital (LUMO) of the free molecule (see Figure 2.14; Ref. 59), which were calculated using the first-principles plane-wave DFT programme DACAPO.[60] How an STM tip can be functionalized to enhance the imaging properties is well known,[61,62] but in general, care has to be taken in the interpretation of images that are acquired with a molecule-terminated tip. In the present case of pentacene on NaCl, the resolution in the images acquired with a metallic tip apex (metal tip) is sufficient to verify that a pentacene molecule at the tip does not introduce additional features (apart from the common phenomenon of so-called double-tip effects) in the nodal plane structures of the orbitals.

STM images showing an intramolecular structure that resembles the LUMO of a molecule have also been obtained for molecules directly adsorbed on metal surfaces at high tunnelling currents.[63–65] However, the direct

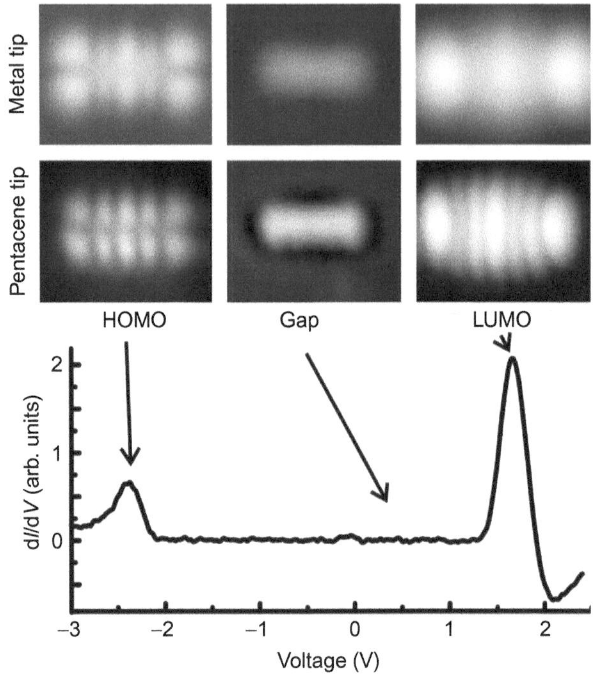

FIGURE 2.13 Constant current STM images taken with a metal tip (upper row) and a pentacene-terminated tip (lower row) at bias voltages corresponding to the peaks and gap region seen in d*I*/d*V* spectroscopy at the centre of a pentacene molecule on NaCl. The differential conductance curve exhibits two peaks that can be attributed to the negative and the positive ion-resonance states, and a broad gap region of low conductance between the peaks. Adapted from Ref. 12 (from left to right, upper row: 2300 mV, 15 pA; 429 mV, 28 pA; 1700 mV, 15 pA. Lower row: 2523 mV, 22 pA; 500 mV, 22 pA; and 1793 mV, 9.6 pA).

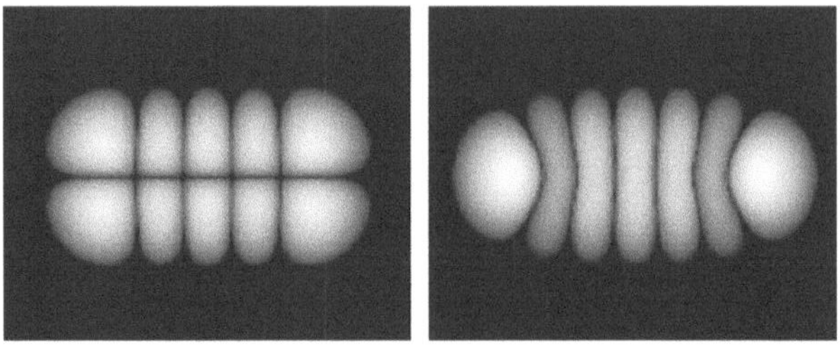

FIGURE 2.14 Constant orbital probability distribution contours ($|\Psi|^2 = 5\times10^{-4}$ Å^{-3}) of the HOMO (left) and the LUMO (right) of the free molecule calculated using the DFT programme DACAPO. Adapted from Ref. 12.

interpretation of these images in terms of the orbital structure is not straightforward. If a flat, π-conjugated molecule is adsorbed onto a metal surface, its electronic coupling to the surface can be such that many molecular resonances contribute to the tunnelling at a given bias voltage. In addition, different through-space and through-molecule tunnelling paths compete with each other and lead to interference effects.[65,66] In the present case of pentacene on 2 ML NaCl, tunnelling at biases corresponding to a molecular resonance will increase the tunnelling current by about two orders of magnitude. This indicates that the through-molecule tunnelling path clearly dominates tunnelling at resonance, which also enables the direct imaging of the HOMO at negative bias voltage polarity. Only in conjunction with the clear, well-separated peaks in the dI/dV spectroscopy, can the bias-dependent STM images be assigned unambiguously to the molecular resonances. The direct correspondence between these images and the molecular orbitals of the free molecule clearly demonstrates that the electronic decoupling provided by the insulating film is sufficient to preserve the inherent electronic properties of the free molecule.

As the work function of NaCl/Cu(111) is known to be about 4 eV,[51] the peaks in the dI/dV spectrum correspond to a position of the HOMO and LUMO of 6.4 and 2.3 eV below the vacuum level, respectively. Note that the lifetime of an electron or hole on the pentacene molecule is strongly extended by the presence of the insulating layer. An electron or hole has to localize in the molecule before it can tunnel away into the substrate. Thus, the situation is very similar to that of electron transport within a molecular solid, where the transport gap differs from the optical HOMO–LUMO gap of the neutral molecule.[67] The peak positions correspond to the NIR and positive ion-resonance states of the adsorbed molecule, which are related to the ionization energy of 6.61 eV[68,69] and the electron affinity of 1.35 eV[70] of the free molecule. These energies are, however, shifted towards the Fermi level because of the electronic polarizations of the NaCl film and the underlying metal.[71] The far slower ionic polarization of the substrate will take place only after the electron or hole has fully localized on the molecule and therefore will not decrease the gap size.

By increasing the thickness of the NaCl layer, the electronic polarization of the metal substrate (image-charge effects) can be reduced and directly correlated to the observed width of the gap, which is 3.3, 4.1, and 4.4 eV for one, two, and three NaCl layers, respectively. When changing the metal substrate orientation from Cu(111) to Cu(100), an almost identical gap width for a bilayer of NaCl is observed, but peak positions shift by 0.35 eV towards lower voltages. This compares nicely with the difference in work functions, which for the clean surfaces is $\Delta\Phi = 0.35$ eV.[72] This observation suggests that the molecular levels of the pentacene molecule are coupled to the vacuum level of the supporting NaCl/Cu system. This is another indication of the electronic decoupling provided by the NaCl film: for π-conjugated molecular films directly adsorbed onto a metal surface, a variety of interface phenomena can be observed that cause the breakdown of the vacuum-level alignment.[73] In contrast, for weakly interacting adsorbates, such as Xe atoms, a vacuum-level alignment is observed and can even be used to measure the local work function of many metallic substrates.[74]

Apart from the peak position, also the peak width is important. The conductance spectra exhibit broad peaks of full width at half maximum (FWHM) of roughly 0.3 eV, although the lifetime of an electron or hole in the molecule should be dramatically increased by the presence of the insulator. For two layers of NaCl, the electronic coupling to the substrate should be so weak that an intrinsic linewidth on the order of few millielectron volts or even less could be expected.[12] The FWHM does not change from two to three layers of NaCl, which also speaks against an FWHM limited by lifetime broadening. The intrinsic energy resolution of elastic STS at 5 K is also in the range of only few millielectron volts. Therefore, the peak-broadening mechanism was investigated by analysing differential conductance spectra of localized states in Cl vacancies in more detail.[47] It turned out that the observed broad line shapes are caused by a strong coupling between the localized electronic state corresponding to the peak in the spectrum and the optical phonons in the film. The underlying physical mechanism can be understood from the Franck–Condon picture illustrated in Figure 2.15 and described in the following. At resonance, the electrons of the STS current temporarily charge the defect or adsorbate. Similar to what could be observed for the charge bistability, the charged ions in the polar insulator react strongly to the temporary addition of charge. In the Franck–Condon principle, this corresponds to a strong displacement of the potential minimum along the axis of ionic displacement. At low temperatures, the vertical electron attachment, which occurs repeatedly when acquiring a spectrum, will encounter the system primarily in the phononic ground state. The slope of the potential at NIR will translate the uncertainty in the positions of the ion cores in the ground state (zero-point motion) into an uncertainty of the energy of the NIR.[47,75,76] To perform a quantitative analysis, the relevant parameters were obtained from DFT calculations, in which the temporary occupation of the electronic state that occurs in the experiment was accounted for. The resulting FWHM agrees well with the experimentally observed one. This simple model applies not only to the vacancies but also to the molecules. In addition, it predicts decreasing

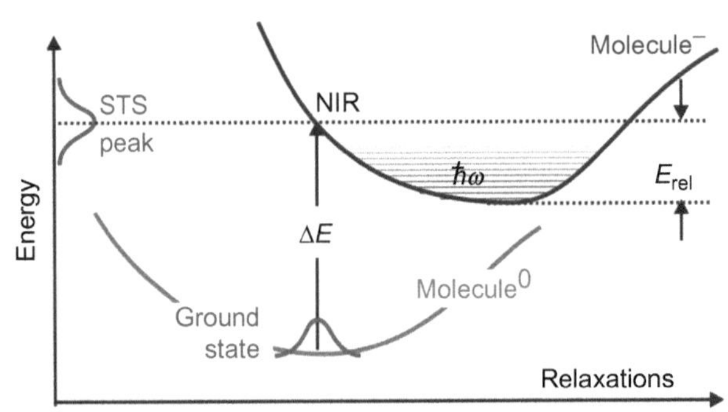

FIGURE 2.15 Illustration of the peak-broadening mechanism due to an electron attachment process based on the Franck–Condon principle.

FWHMs for different insulators in the order from LiF to NaCl, RbI, and Xe. This prediction was confirmed for pentacene molecules in experiments on all these films.[47] These findings have strong implications for molecular electronics. If molecular electronics is to be extended to more complex structures of connected molecular devices, rather than just a single molecule between two metallic leads, every molecular device will be embedded in a specific environment, which, as is described above, may have dramatic effects on the conductance properties even if it is completely insulating. However, the strong phonon broadening may also be used to one's advantage. For example, it may help the electron transport between two levels that are not perfectly aligned.

6. MANIPULATION OF MOLECULES: MOLECULAR/ORBITAL ENGINEERING

The ability of the STM to induce and control chemical reactions on the single-molecule level has been one of the most exciting prospects of molecular manipulation. The controlled breaking/forming of individual chemical bonds proceeds via the nonthermal vibrational and/or electronic excitation of molecules. The voltage and current dependence of the observed reaction rate provides direct information on the underlying excitation process, that is, for example, whether single or multiple electron excitation prevails and which vibrational modes are involved. While a large number of these studies exist on metal substrates,[77–80] on insulating substrates, similar manipulation-based studies on single-molecule chemistry are still very rare.

In most experiments of STM-induced chemical reactions, inelastic electron tunnelling (IET) is employed. In particular, in the context of insulating films, it is important to distinguish different types of these IET processes. On metal substrates, the electronic local density of states (LDOS) of an adsorbate is usually broadened in energy and has a finite value close to the Fermi level of the substrate. In this case, IET may occur as soon as the bias voltage times the elementary charge exceeds the energy of the elementary excitation, that is, $h\nu$ of the vibration. This is what IET commonly refers to on metal substrates. A different type of IET may occur, when electrons can resonantly tunnel into vibronic states, which are observed as satellites beyond electronic resonances.[6] This IET process occurs at much higher voltages and does not require a finite LDOS near the Fermi level.

On an insulating film, the latter type of IET is very efficient because of the longer electronic lifetimes.[6,10,12] However, as the insulating film may not only reduce the LDOS of a molecule at the Fermi level but may also alter the lifetime of an vibrational excitation, also the former process apparently can still be efficient, as is discussed in the following paragraph for a small molecule, namely water.

For large molecules with a delocalized π-electronic system, manipulation on insulating films offers the combination of chemical bond breaking/formation and orbital imaging, that is, orbital engineering. Following the discussion of the dissociation of water molecules, we will present in detail three examples of how to controllably change the orbital structure by molecular manipulation. The first two of these examples discuss bond formation between a metal atom and a molecule to form metal molecule complexes. The third example discusses the realization of a unique planar molecular switch based on tunnelling-induced hydrogen tautomerization.

6.1. Manipulation of a Small Molecule: Water Dissociation on MgO Films

One very important aspect of molecular manipulation on insulating films is related to the catalytic reactivity of oxides. Even though MgO is catalytically rather inactive, the STM-induced dissociation of water[81] is a first example of and an important step forward in single-molecule chemistry of small molecules on oxides. In addition, the dissociation of water is of high technical importance as a candidate for generating hydrogen as a source for clean energy production. Interestingly, the possibility of spontaneous dissociation of water on the MgO surface has been disputed in the literature. To provide atomic-scale insights in this issue, Shin *et al.* have investigated the dissociation of single water molecules on a two atomic layer thick magnesium oxide film. Depending on the tunnelling voltage, the authors could select two different excitation mechanisms to dissociate the H_2O molecule. At low bias voltage and small tunnelling currents, first lateral hopping of the molecule could be induced by excitation of the H_2O scissor (196 meV) and asymmetric O—H stretch mode (448 meV). Further increase of the tunnelling current at the energy of the O—H stretch mode resulted in multiple vibrational excitation of this mode and finally in the dissociation of the molecule into the OH^- and hydrogen. The relatively small bias voltages as compared to electronic excitations of the molecule show that it is possible to vibrationally excite the molecule without the involvement of vibronic levels also on an insulating film. The calculated dissociation barrier for this process on the MgO film was only 0.77 eV compared to 1.07 eV on bulk MgO, that is, that the substrate is an ultrathin insulating film is essential. By increasing the electron energy to 1.5 eV, that is, in the range of calculated LUMO of H_2O, direct electronic excitation was possible resulting in the full dissociation of the water molecule into oxygen and hydrogen. The demonstrated ability to select the excitation mechanism (vibrational, electronic) mediates therefore control over the chemical reaction route and therefore the reaction products (OH,O).

6.2. Organometallic Synthesis: Au–Pentacene

Here, we discuss the bond formation between a single metal atom (Au) and a molecule (pentacene) to build a Au–pentacene complex.[13] This example also constitutes the first step in the electrical contacting of a molecular device with atomic precision. Such atomic control of the contact formation would allow a selective coupling to molecular orbitals including their phase, which may be used to implement new device functions.[82] As discussed, single-molecule chemistry by means of STM on an insulating surface differs from the well-established manipulation on metal substrates, on which single-molecule synthesis can be split into two distinct steps[62,78,83]: on a flat metal surface, the first step, in which the reactants are brought close to each other by lateral manipulation,[1] is relatively easy because of the small ratio of diffusion barrier to binding energy. In the second, more complex bond-formation step, much higher energy barriers have to be overcome. Therefore, this step is initiated by nonthermal excitation of the reactants using IET. In the case of an insulating substrate, the ratio of diffusion barrier to binding energy is much larger so that excitation by IET is already required to induce the lateral movement. In contrast to common inelastic electron tunnelling spectroscopy (IETS) of molecules on metal surfaces, IET excitation on insulating films proceeds via tunnelling through vibronic levels (see above), which is a much more efficient process because of the longer electronic lifetimes.[6,10,12] The actual bond formation between the metal atom and the molecule, however, requires a smaller activation energy, such that the IET excitation process applied to achieve lateral movement is already sufficient to also initiate the synthesis. Figure 2.16 shows the bond formation between a pentacene molecule and a gold atom on a bilayer of NaCl. In Figure 2.16A, the reactants are already located close to each other. The bond was formed by IET into the LUMO of pentacene at a voltage of 1.75 V and a current of only a few picoampere. The resulting complex (Figure 2.16B) has a mirror plane perpendicular to the long axis of the molecule, indicating that the gold atom is attached to the central ring of pentacene (6-gold–pentacene). Using IET, the bond can be broken again (Figure 2.16C), and in a subsequent step, a new bond can be formed with the gold attached in a new position (Figure 2.16D): the dark feature in the STM image of the complex indicates that the gold atom is slightly off-centre (5-gold–pentacene). The complete reversibility of the complex formation suggests that this complex formation is an addition reaction of the gold atom to one of the pentacenes aromatic rings and, surprisingly, involves neither the substitution of a hydrogen atom nor a defect creation in the substrate.[13]

The differential conductance signal (dI/dV) acquired with the STM tip above the 6-gold–pentacene complex in Figure 2.17D exhibits two pronounced peaks at -1.5 and $+1.2$ V separated by a broad gap. As discussed in the previous section for isolated pentacene, images taken at bias voltages that correspond to the gap region (in-gap conditions) and to the peaks in dI/dV spectra (resonance conditions) differ greatly. As can be seen in Figure 2.17, these resonance-condition images, A and C, show pronounced intramolecular resolution having a large corrugation of 2.0 and 2.6 Å, respectively. In contrast, for in-gap conditions, the apparent height measures only 1.0 Å (Figure 2.17B). The orbital structure of the complex is delocalized over the

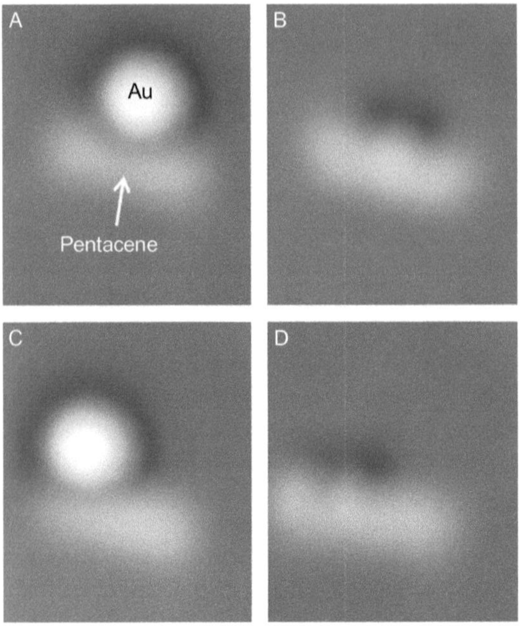

FIGURE 2.16 Making and breaking of a chemical bond between a single pentacene molecule and gold atom on an NaCl bilayer supported by a Cu(100) substrate. The STM image (A) shows the molecule and the gold atom prior to bond formation. Image (B) shows a molecule–metal complex (6-gold–pentacene) after resonant tunnelling through the LUMO of the pentacene molecule. The metal atom can be detached (C) and reattached (D) in a different position, forming another structural isomer of the complex (5-gold–pentacene). All images are 26 Å × 22 Å in size and taken at a bias of +0.34 V. Adapted from Ref. 13.

entire complex, including the gold atom, indicating a covalent bond between the pentacene molecule and the gold atom. When the gold atom is very close to a pentacene molecule but not bound to it, the corresponding STM image exhibits no sign of bonding or common orbital structure, but is merely a superposition of the unperturbed LUMO of the pentacene molecule and the protrusion of the gold atom.

The charge state of the 6-gold–pentacene complex was found to be neutral, as inferred from the absence of scattering of the NaCl/Cu(111) IS electrons at the complex as described above and in Ref. 12. Thus, the complex accommodates an odd number of electrons and has a singly occupied molecular orbital (SOMO) near the Fermi level. This explains why the orbital structure looks almost the same for both bias voltage polarities (see Figure 2.17A and C): in both cases, the electrons tunnel through the same orbital, either temporarily emptying it ($V<0$) or temporarily filling it with a second electron ($V>0$). Consequently, the broad gap in the dI/dV spectra of the complexes is not a HOMO–LUMO gap but solely due to the Coulomb energy associated with adding or removing an electron to or from the same orbital of the complex. Thereby, the separation of the two peaks labelled SOMO provides a simple and direct measure of this important parameter. Note that for an isolated pentacene, the HOMO–LUMO peak separation is much larger (4.1 eV; Ref. 12). This value can be regarded as the sum of the energy needed to excite an electron from the HOMO to the LUMO plus a similar Coulomb energy associated with adding or removing an electron to or from that orbital.

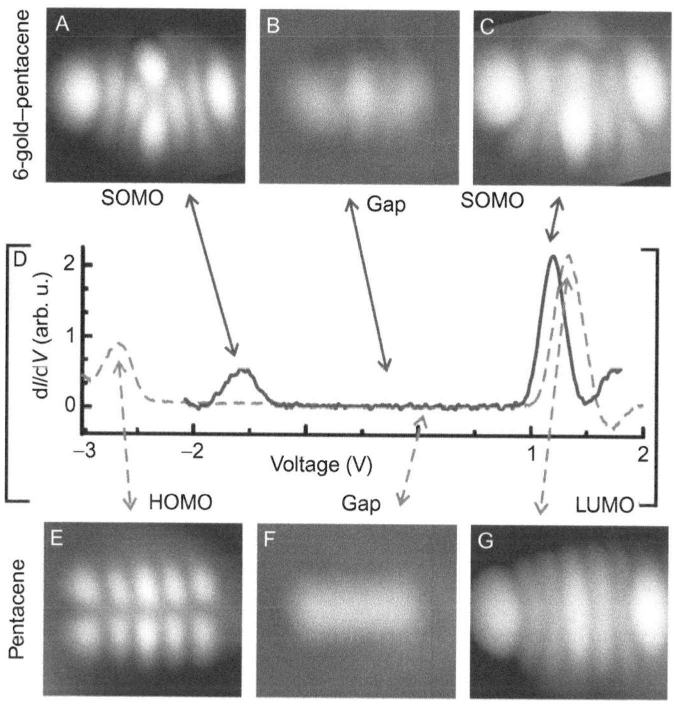

FIGURE 2.17 Bias-dependent STM images and differential conductance (d*I*/d*V*) spectra of a 6-Au-pentacene complex (A–C) and a pentacene molecule (E–G) on the NaCl film. The d*I*/d*V* spectrum of the 6-gold–pentacene complex exhibits two distinct peaks at −1.5 and +1.2 V (solid line). For comparison, the plot also shows the d*I*/d*V* signal of an isolated pentacene molecule (dashed line). Adapted from Ref. 13.

Of particular interest is the nature of the Au–pentacene as identified by DTF calculations.[13] In the case 6-gold–pentacene, the Au atom is bonded to both a Cl anion and a C atom in the central ring of the molecule. The H atom is tilted upwards so that the C atom is in a nearly tetrahedral bond configuration with respect to its four neighbours, the Au atom, the H atom, and the two adjacent C atoms, indicating a sp_2-to-sp_3 rehybridization of the C atom. The bonding of the Au atom to the pentacene molecule results in sizeable relaxations in the NaCl film but only for the Cl anion bonded to the Au atom and the Na cation below this anion. The addition of an Au atom in its neutral state to the pentacene molecule results in a radical complex with an odd number of electrons, with predominant orbital interactions among their frontier orbitals. The HOMO and LUMO of the molecule and the Au(6s) state interact and form three new orbitals whose characters can be rationalized in a simple three-state model. The HOMO of the complex is predominantly a bonding combination of the Au(6s) state with the HOMO of pentacene, whereas the LUMO is an antibonding combination with the pentacenes LUMO. The remaining SOMO is a linear combination of the Au(6s) state with both the HOMO (with antibonding character) and the LUMO (with bonding character). Further, it was found that the addition of the Au atom to the molecule is electrophilic as there is a small net electron transfer to Au due to the larger electronegativity of the Au atom than of the pentacene molecule. Au–pentacene constitutes a simple example of 'orbital engineering', because by creating different isomers, both the nodal structure itself and the relative weight of the probability distribution in different parts of the molecule can be controlled. As the covalent bond involves the delocalized electronic orbitals of the pentacene molecule, it also facilitates a strong coupling of the electronic states that are involved in electron transport.

6.3. Metal–Ligand Complex Formation

Coordination chemistry provides an alternative way of synthesizing larger molecular structures that is based on the bonding of organic ligands to metal atom centres. The large number of different metal–ligand pairs with different bonding strengths and coordination numbers offers great flexibility for organizing single molecular building blocks into supramolecular structures. Therefore, metal–ligand-based bonding is also one of the key strategies in self-assembly on metal surfaces.[84–86] A step-by-step synthesis of single metal–ligand complexes by molecular manipulation and direct visualization of the frontier molecular orbitals, resulting from electronic decoupling by an insulating film substrate, will enable a deeper understanding of these systems. In the following, we discuss the formation of a linear complex consisting of a transition-metal atom (Fe, Ni) and two 9,10-dicyanoanthracene (DCA) molecules.[87]

The first step of the synthesis of an $Fe(DCA)_2$ complex is to attach an Fe atom to a DCA molecule by moving an Fe atom close to a DCA molecule with lateral manipulation and attaching it by means of IET excitation at a bias voltage of 0.3–0.4 V. To complete the $M(DCA)_2$ complex, the second DCA molecule is moved close to and attached to the complex using IET excitation (bias voltage of 1.5–2.0 V and current of 5 pA). The manipulation steps were carried out using a molecule-terminated tip to prevent pickup of the molecule to be manipulated. After each manipulation step, STM imaging was used to determine the exact molecular geometry.

Using STM and STS, two resonances were observed for the $Fe(DCA)_2$ complex when the tip was located above the DCA ligands at bias voltages of 1.16 and −0.76 V. These two frontier orbitals stem from the LUMO of the isolated DCA and are delocalized over the entire $Fe(DCA)_2$ complex. Although the STM images of the two orbitals appear similar, they do not exactly have the same structure, as the image at negative bias exhibits significantly more intensity in the centre of the molecule. Apart from the delocalized states, also a higher-lying molecular resonance at 1.67 V localized on the metal atom is observed, which suggests that this resonance derives from the 4s and/or the partially filled 3d states of the iron atom.

Although the changes in the orbital structure of $Fe(DCA)_2$ when compared with a single DCA ligand clearly show the formation of a metal–ligand complex, a clear assignment of these resonances to specific orbitals could only be done by comparison with detailed DFT calculations carried out on an isolated linear $Fe(DCA)_2$ complex without including the substrate. The best agreement between theory and experiment was found for a neutral complex. In the complex, the calculated iron-nitrogen bond strength was 1.1 eV. Most importantly, the ground state of $Fe(DCA)_2$ has four unpaired electrons, that is, a quintet spin state, of which three are strongly localized on the Fe atom and show a large spin splitting of 7–8 eV. The fourth unpaired electron is delocalized over the entire complex, and the DFT calculations predict that the SOMO exhibits a spin splitting of almost 1 eV, suggesting a substantial exchange coupling in this delocalized electron system. Apart from the SOMO, a spin-split LUMO exists that is also delocalized over the entire complex. In addition, the calculations predict the β-resonance of another SOMO localized on the metal atom at an energy that is only slightly higher than the LUMO. To be consistent with the experimental findings, the first measured resonance at positive bias has to involve tunnelling through the unoccupied delocalized SOMO and LUMO levels. The spin state and the order and spatial symmetries of the molecular orbitals of such a complex in general depend strongly on the nature of the metal centre. For example, similar work on $Ni(DCA)_2$ complexes favours a spin singlet state.[87] These results demonstrate the possibility of using atomic manipulation to build a molecule-metal-molecule bridge[88] with strong electronic coupling between the molecular parts and, as suggested by the DFT calculations, a large exchange coupling in a delocalized π-electron system, which could be important for the design of novel spin-based molecular electronic devices.

6.4. Molecular Switching Based on a Tautomerization Reaction

Next, we describe a molecular switch based on a tautomerization reaction.[15] The switch consists of a free-base naphthalocyanine molecule, which hosts two hydrogen atoms in its central cavity. The hydrogens reduce the otherwise fourfold rotational symmetry (D_{4h}) of the molecule to a twofold symmetry (D_{2h}). Therefore, there are two equivalent positions of the two hydrogens. As in the case of pentacene, the frontier molecular orbitals can be directly imaged.

STM images acquired with a molecule-terminated tip at bias voltages corresponding to the HOMO, LUMO and to the in-gap condition are reproduced in Figure 2.18. The molecule is adsorbed along the nonpolar direction of the underlying NaCl bilayer film on Cu(111). As in the case of pentacene, the images acquired at voltages corresponding to molecular resonances compare very well with the calculated contours of constant density of the electronic wave functions (DFT; Ref. 15) of a free molecule as shown in the

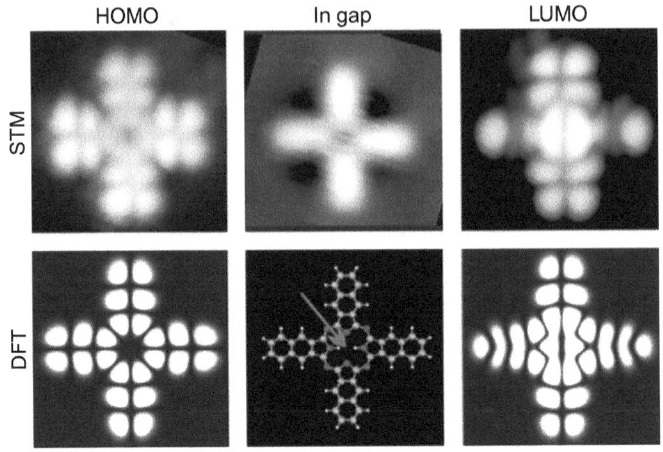

FIGURE 2.18 STM images at the HOMO, LUMO, and 'in-gap' bias of naphthalocyanine (upper row) and comparison with the calculated orbital images of the free molecule (lower row). The centre panel in the lower row shows the structure model to scale. Adapted from Ref. 15.

lower part of Figure 2.18. Whereas the HOMO image has fourfold symmetry, the STM image corresponding to the LUMO has a twofold rotational symmetry. Therefore, the LUMO image allows an easy determination of the position of the imino hydrogens: the arms with the hydrogens exhibit a single lobe structure at the end, whereas along the other two arms, there is a nodal plane.

Most interestingly, the molecules can be switched in a controlled fashion between the two possible hydrogen positions using inelastic tunnelling. The hydrogen tautomerization can be induced by positioning the tip somewhere above the molecule and increasing the bias to a value significantly above the LUMO resonance. Because the LUMO images differ distinctly for the two tautomers, the reaction can be directly monitored in the current signal or the vertical tip position in constant-height or constant-current time traces, respectively. In these traces, the current or vertical tip position switches back and forth between two well-defined levels as shown in the left panel of Figure 2.19 for a bias voltage of 1.7 V. When the bias is lowered to 0.8 eV, the LUMO can be imaged and it can be seen that the two current levels correspond to a 90° rotation of the orientation of the LUMO. Using DFT calculations, this change can be assigned to changes in the position of the imino hydrogens in the central cavity, that is, hydrogen tautomerization (see centre images of Figure 2.19). A rotation of the entire molecule can be ruled out because also a switching of molecules at step edges and in arrays of molecules is observed, where no rotation of the molecule can occur. Similarly, also the switching of a molecule on a hexagonally arranged Xe monolayer shows only two orientations.

The dependence of the switching rate on the current is linear, and the distribution of residence times in the low- and the high-current state is stochastic.

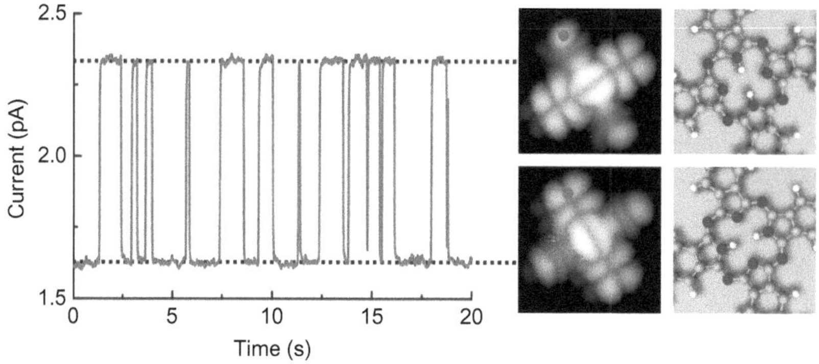

FIGURE 2.19 Tunnelling current-induced switching of a single naphthalocyanine molecule. (Left) Current signal recorded at a bias of 1.7 V with the tip positioned above the end of the molecule (red dot in STM images). (Centre) Orbital images showing the two orientations of the LUMO (2 pA, 0.7 V). (Right) Schematics of the corresponding hydrogen positions in the central cavity of the molecule. Adapted from Ref. 15. (For interpretation of the references to colour in this figure legend, the reader is referred to the Web version of this chapter.)

These facts are consistent with a one-electron process. The switching rate increases with increasing bias voltage in a roughly exponential fashion. Because of the reliability of the switching process, the spatial dependence of the switching rate, that is, its dependence on the position of the electron injection into the molecule, can be measured. Interestingly, the switching rate is different for the two arms and largest when the tip is above the far periphery of the molecule, that is, > 10 Å from the centre of the molecule. This is in contrast to the typical IET mechanism.[78,80,83] The lifetime of an additional electron in a molecular resonance is expected to be relatively long[6,10,13] because of the decoupling provided by the insulating film. When the molecules were directly adsorbed on a metal substrate (e.g. Cu(100)), no evidence of this kind of tautomerization reaction was observed. This highlights the role of electron and/or energy transport within the molecule, making this system particularly interesting for such transport studies.[89,90] The tautomerization reaction can be regarded as a local sensor, which, together with local current injection, enables transport studies in a two-terminal configuration. This opens up the possibility of studying intermolecular energy and electron transport. To this end, three naphthalocyanines were moved very close to each other by lateral manipulation with an STM as shown in Figure 2.20A–D. In this configuration, it was possible to reversibly switch one molecule by current injection through the neighbouring molecules. Because energy transport in the absence of electron transport (by dipolar or other mechanisms) across the molecules seems extremely unlikely, the switching yield in this experiment directly relates to the electron transport properties through the molecule, into which the current is injected, the coupling of the two adjacent molecules, and the properties of the sensing molecule. This class of molecules is also well suited for use in self-assembled monolayers

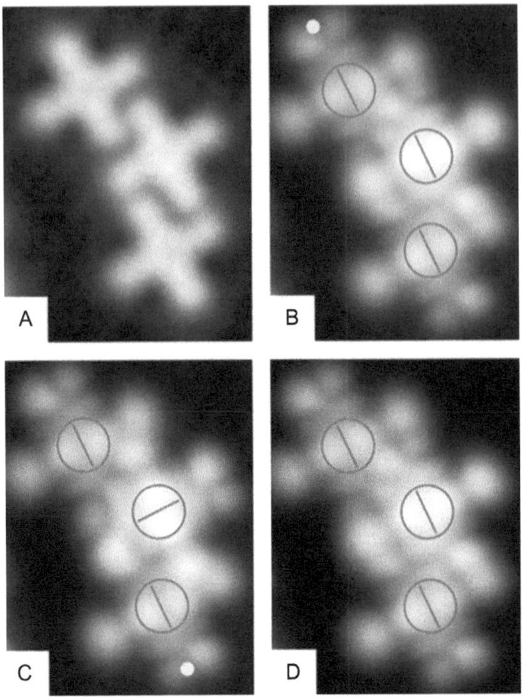

FIGURE 2.20 Examples of interacting and non-interacting assemblies of molecules. (A) A trimer of naphthalocyanine molecules on NaCl bilayer formed by STM manipulation. Current injection through the top or bottom molecules of the trimer, marked by yellow dots in (B) and (C), can cause the switching of the molecule in the middle as shown by the LUMO images. Images are 66 × 66 Å. Adapted from Ref. 15. (For interpretation of the references to colour in this figure legend, the reader is referred to the Web version of this chapter.)

because they are planar and the switching does not involve conformational changes at the periphery of the molecule. It is anticipated that the switching will also work with molecules embedded in all solid-state devices and in multicore porphyrin-class molecules acting as more complex devices, which might make them potentially useful as nonvolatile memory with extremely high density, as has often been proposed.[91–94]

7. CONCLUSION

Atomic and molecular manipulation on ultrathin insulating films open up new possibilities for controlling matter on the atomic scale, with control of the charge state of single metal atoms and orbital imaging/engineering being key techniques. These processes are based on the weak electronic coupling to a conducting substrate and thus a significantly increased lifetime of the tunnelling electron on the adsorbate. This gives rise to increased electron phonon interaction and therefore enhanced efficiency of bond breaking/formation processes driven by inelastic tunnelling, which in combination with orbital imaging open up new possibilities in single-molecule chemistry. As atomically perfect insulating films with varying thickness and of different material can be grown almost routinely, substrate-related properties can be conveniently studied and exploited. This provides deeper insight and understanding of adsorbate insulator interaction, a key aspect in surface and interface science, not only in the space but also in the time domain.[9] Whereas with STM such studies are limited to films of only a few atomic layers, AFM will allow their extension to thicker films without losing atomic resolution.[95,96] This constitutes a first step in the investigation of new electronic transport phenomena in planar molecular networks, opening up new opportunities in monomolecular electronics.[97]

REFERENCES

1. Eigler DM, Schweizer EK. Positioning single atoms with a scanning tunnelling microscope. *Nature* 1990;**344**(6266):524–6.
2. Ho W. Single-molecule chemistry. *J Chem Phys* 2002;**117**(24):11033–61.
3. Mayne AJ, Dujardin G, Comtet G, Riedel D. Electronic control of single-molecule dynamics. *Chem Rev* 2006;**106**(10):4355–78.
4. Cavar E, Blueml M-C, Pivetta M, Patthey F, Chergui M, Schneider W-D. Fluorescence and phosphorescence from individual C_{60} molecules excited by local electron tunneling. *Phys Rev Lett* 2005;**95**(19):196102.
5. Qiu XH, Nazin GV, Ho W. Vibrationally resolved fluorescence excited with submolecular precision. *Science* 2003;**299**(5606):542–6.
6. Qiu XH, Nazin GV, Ho W. Vibronic states in single molecule electron transport. *Phys Rev Lett* 2004;**92**(20):206102.
7. Heinrich AJ, Gupta JA, Lutz CP, Eigler DM. Single-atom spin-flip spectroscopy. *Science* 2004;**306**(5695):466–9.
8. Hirjibehedin CF, Lutz CP, Heinrich AJ. Spin coupling in engineered atomic structures. *Science* 2006;**312**:1021–4.
9. Loth S, Etzkorn M, Lutz CP, Eigler DM, Heinrich AJ. Measurement of fast electron spin relaxation times with atomic resolution. *Science* 2010;**329**(5999):1628–30.
10. Repp J, Meyer G, Rieder K-H. Snell's law for surface electrons: refraction of an electron gas imaged in real space. *Phys Rev Lett* 2004;**92**(3):036803.
11. Wu SW, Ogawa N, Nazin GV, Ho W. Conductance hysteresis and switching in a single-molecule junction. *J Phys Chem C* 2008;**112**:5241.
12. Repp J, Meyer G, Paavilainen S, Olsson FE, Persson M. Scanning tunneling spectroscopy of Cl vacancies in NaCl films: strong electron-phonon coupling in double-barrier tunneling junctions. *Phys Rev Lett* 2005;**22**:225503.
13. Repp J, Meyer G, Paavilainen S, Olsson FE, Persson M. Imaging bond formation between a gold atom and pentacene on an insulating surface. *Science* 2006;**312**(5777):1196–9.
14. Wu SW, Nazin GV, Chen X, Ho W. Control of relative tunneling rates in single molecule bipolar electron transport. *Phys Rev Lett* 2004;**93**(23):236802.
15. Liljeroth P, Repp J, Meyer G. Current-induced hydrogen tautomerization and conductance switching of naphthalocyanine molecules. *Science* 2007;**317**(5842):1203–6.
16. Freund HJ, Pacchioni G. Oxide ultra-thin films on metals: new materials for the design of supported metal catalysts. *Chem Soc Rev* 2008;**37**(10):2224–42.
17. Netzer FP, Allegretti F, Surnev S. Low-dimensional oxide nanostructures on metals: hybrid systems with novel properties. *J Vac Sci Technol B* 2010;**28**(1):1–16.
18. Libuda J, Winkelmann F, Bäumer M, Freund HJ, Bertrams T, Neddermeyer H, et al. Structure and defects of an ordered alumina film on NiAl(110). *Surf Sci* 1994;**318**(1–2):61–73.

19. Kresse G, Schmid M, Napetschnig E, Shishkin M, Kohler L, Varga P. Structure of the ultrathin aluminum oxide film on NiAl(110). *Science* 2005;**308**(5727):1440–2.
20. Tsukahara N, Noto K-i, Ohara M, Shiraki S, Takagi N, Takata Y, et al. Adsorption-induced switching of magnetic anisotropy in a single iron(II) phthalocyanine molecule on an oxidized Cu(110) surface. *Phys Rev Lett* 2009;**102**(16):167203.
21. Fu YS, Ji SH, Zhang T, Chen X, Ma X-C, Jia J-F, et al. Ultrathin lead oxide film on Pb(111) and its application in single spin detection. *Appl Phys Lett* 2009;**95**(6):063107–13.
22. Schintke S, Messerli S, Pivetta M, Patthey F, Libioulle L, Stengel M, et al. Insulator at the ultrathin limit: MgO on Ag(001). *Phys Rev Lett* 2001;**87**(27):276801.
23. Avouris P, Wolkow R. Scanning tunneling microscopy of insulators: CaF_2 epitaxy on Si (111). *Appl Phys Lett* 1989;**55**(11):1074–6.
24. Klust A, Yu Q, Olmstead MA, Ohta T, Ohuchi FS, Bierkandt M, et al. Contrast in scanning probe microscopy images of ultrathin insulator films. *Appl Phys Lett* 2006;**88**(6):063107.
25. Viernow J, Petrovykh DY, Kirakosian A, Lin J-F, Men FK, Henzler M, et al. Chemical imaging of insulators by STM. *Phys Rev B* 1999;**59**(15):10356.
26. Hebenstreit W, Redinger J, Horozova Z, Schmid M, Podlucky R, Varga P. Atomic resolution by STM on ultra-thin films of alkali halides: experiment and local density calculations. *Surf Sci* 1999;**424**:L321.
27. Bennewitz R, Barwich V, Bammerlin M, Loppacher C, Guggisberg M, Baratoff A, et al. Ultrathin films of NaCl on Cu(111): a LEED and dynamic force microscopy study. *Surf Sci* 1999;**438**(1–3):289–96.
28. Fölsch S, Helms A, Zöphel S, Repp J, Meyer G, Rieder K-H. Self-organized patterning of an insulator-on-metal system by surface faceting and selective growth: NaCl/Cu(211). *Phys Rev Lett* 2000;**84**(1):123–6.
29. Kramer J, Tegenkamp C, Pfnur H. The growth of NaCl on flat and stepped silver surfaces. *J Phys Condens Matter* 2003;**15**(38):6473–83.
30. Repp J, Fölsch S, Meyer G, Rieder K-H. Ionic films on vicinal metal surfaces: enhanced binding due to charge modulation. *Phys Rev Lett* 2001;**86**(2):252–5.
31. Repp J, Meyer G, Olsson FE, Persson M. Controlling the charge state of individual gold adatoms. *Science* 2004;**305**(5683):493–5.
32. Tu XW, Mikaelian G, Ho W. Controlling single-molecule negative differential resistance in a double-barrier tunnel junction. *Phys Rev Lett* 2008;**100**(12):126807.
33. Fölsch S, Barjenbruch U, Henzler M. Atomically thin epitaxial films of NaCl on germanium. *Thin Solid Films* 1989;**172**(1):123–32.
34. Glöckler K, Sokolowski M, Soukopp A, Umbach E. Initial growth of insulating overlayers of NaCl on Ge(100) observed by scanning tunneling microscopy with atomic resolution. *Phys Rev B* 1996;**54**(11):7705.
35. Schwennicke C, Schimmelpfennig J, Pfnür H. Morphology of thin NaCl films grown epitaxially on Ge(100). *Surf Sci* 1993;**293**(1–2):57–66.
36. Such B, Goryl G, Godlewski S, JKolodziej JJ, Szymonski M. PTCDA molecules on a KBr/InSb system: a low temperature STM study. *Nanotechnology* 2008;**19**(47):475705.
37. Gross L, Schlittler RR, Meyer G, Fendt LA, Diederich F, Glatzel T, et al. Contacting self-ordered molecular wires by nanostencil lithography. *J Vac Sci Technol B* 2010;**28**(3): C4D34–C4D39.
38. Dong ZC, Guo XL, Trifonov AS, Dorozhkin PS, Miki K, Kimura K, et al. Vibrationally resolved fluorescence from organic molecules near metal surfaces in a scanning tunneling microscope. *Phys Rev Lett* 2004;**92**(8):086801.

39. Franke KJ, Schulze G, Henningsen N, Fernandez-Torrente I, Pascual JI, Zarwell S, et al. Reducing the molecule-substrate coupling in C_{60}-based nanostructures by molecular interactions. *Phys Rev Lett* 2008;**100**(3):036807.
40. Repp J, Meyer G. Scanning tunneling spectroscopy of molecules on insulating films. *Chimia* 2010;**64**:370–5.
41. Smoluchowski R. Anisotropy of the electronic work function of metals. *Phys Rev* 1941;**60**(9):661.
42. Olsson FE, Persson M. A density functional study of adsorption of sodium-chloride overlayers on a stepped and a flat copper surface. *Surf Sci* 2003;**540**(2–3):172–84.
43. Crommie MF, Lutz CP, Eigler DM. Confinement of electrons to quantum corrals on a metal surface. *Science* 1993;**262**(5131):218–20.
44. Hasegawa Y, Avouris P. Direct observation of standing wave formation at surface steps using scanning tunneling spectroscopy. *Phys Rev Lett* 1993;**71**(7):1071–4.
45. Hörmandinger G. Imaging of the Cu(111) surface state in scanning tunneling microscopy. *Phys Rev B* 1994;**49**(19):13897.
46. Li J, Schneider W-D, Berndt R, Bryant OR, Crampin S. Surface-state lifetime measured by scanning tunneling spectroscopy. *Phys Rev Lett* 1998;**81**(20):4464–7.
47. Repp J, Meyer G, Stojkovic SM, Gourdon A, Joachim C. Molecules on insulating films: scanning-tunneling microscopy imaging of individual molecular orbitals. *Phys Rev Lett* 2005;**94**(2):026803.
48. Hotzel A, Moos G, Ishioka K, Wolf M, Ertl G. Femtosecond electron dynamics at adsorbate-metal interfaces and the dielectric continuum model. *Appl Phys B Lasers Opt* 1999;**68**(3):615–22.
49. Smith NV. Phase analysis of image states and surface states associated with nearly-free-electron band gaps. *Phys Rev B* 1985;**32**(6):3549.
50. Forstmann F. On the theory of surface states in nearly free electron systems. *Z Phys* 1970;**235**(1):69–74.
51. Bennewitz R, Bammerlin M, Guggisberg M, Loppacher C, Baratoff A, Meyer E, et al. Aspects of dynamic force microscopy on NaCl/Cu(111): resolution, tip-sample interactions and cantilever oscillation characteristics. *Surf Interface Anal* 1999;**27**:462.
52. Limot L, Pehlke E, Kröger J, Berndt R. Surface-state localization at adatoms. *Phys Rev Lett* 2005;**94**(3):036805.
53. Olsson FE, Persson M, Borisov AG, Gauyacq J-P, Lagoute J, Fölsch S. Localization of the Cu (111) surface state by single cu adatoms. *Phys Rev Lett* 2004;**93**(20):206803.
54. Pacchioni G, Giordano L, Baistrocchi M. Charging of metal atoms on ultrathin MgO/Mo(100) films. *Phys Rev Lett* 2005;**94**(22):226104.
55. Olsson FE, Paavilainen S, Persson M, Repp J, Meyer G. Multiple charge states of Ag atoms on ultrathin NaCl films. *Phys Rev Lett* 2007;**98**(17):176803.
56. Olsson FE, Persson M, Repp J, Meyer G. Scanning tunneling microscopy and spectroscopy of NaCl overlayers on the stepped Cu(311) surface: experimental and theoretical study. *Phys Rev B* 2005;**71**(7):075419.
57. Hirjibehedin CF, Lin CY, Otte AF, Ternes M, Lutz CP, Jones BA, et al. Large magnetic anisotropy of a single atomic spin embedded in a surface molecular network. *Science* 2007;**317**:1199–203.
58. Eigler DM, Lutz CP, Rudge WE. An atomic switch realized with the scanning tunnelling microscope. *Nature* 1991;**352**(6336):600–3.
59. Endres RG, Fong CY, Yang LH, Witte G, Wöll C. Structural and electronic properties of pentacene molecule and molecular pentacene solid. *Comput Mater Sci* 2004;**29**(3):362–70.

60. *CAMP. DACAPO version 2.6.1, using GGA(PW91) a 26 × 17 × 13 Å^3 unit cell, 400 eV plane-wave cut-off energy, and 0.05 eV/Å residual forces.* In: Lyngby, Denmark: CAMP, Technical University of Denmark; 2002. *http://www.fysik.dtu.dk/campos.*
61. Bartels L, Meyer G, Rieder K-H. Controlled vertical manipulation of single CO molecules with the scanning tunneling microscope: a route to chemical contrast. *Appl Phys Lett* 1997;**71**(2):213–5.
62. Hahn JR, Ho W. Oxidation of a single carbon monoxide molecule manipulated and induced with a scanning tunneling microscope. *Phys Rev Lett* 2001;**87**(16):166102.
63. Böhringer M, Schneider W-D, Berndt R. Scanning tunneling microscope-induced molecular motion and its effect on the image formation. *Surf Sci* 1998;**408**(1–3):72–85.
64. Lagoute J, Kanisawa K, Fölsch S. Manipulation and adsorption-site mapping of single pentacene molecules on Cu(111). *Phys Rev B* 2004;**70**(24):245415.
65. Soe WH, Manzano C, De Sarkar A, Chandrasekhar N, Joachim C. Direct observation of molecular orbitals of pentacene physisorbed on Au(111) by scanning tunneling microscope. *Phys Rev Lett* 2009;**102**(17):176102.
66. Sautet P, Bocquet M-L. Shape of molecular adsorbates in STM images: a theoretical study of benzene on Pt(111). *Phys Rev B* 1996;**53**(8):4910.
67. Pope M, Svenberg C. *Electronic processes in organic crystals.* Oxford: Clarendon Press; 1982.
68. Coropceanu V, Malagoli M, da Silva Filho DA, Gruhn NE, Bill TG, Bredas JL. Hole- and electron-vibrational couplings in oligoacene crystals: intramolecular contributions. *Phys Rev Lett* 2002;**89**(27):275503.
69. Schmidt W. Photoelectron spectra of polynuclear aromatics. V. Correlations with ultraviolet absorption spectra in the catacondensed series. *J Chem Phys* 1977;**66**(2):828–45.
70. Crocker L, Wang T, Kebarle P. Electron affinities of some polycyclic aromatic hydrocarbons, obtained from electron-transfer equilibria. *J Am Chem Soc* 1993;**115**(17):7818–22.
71. Halasinski TM, Hudgins DM, Salama F, Allamandola LJ, Bally T. Electronic absorption spectra of neutral pentacene (C22H14) and its positive and negative ions in Ne, Ar, and Kr matrices. *J Phys Chem A* 2000;**104**(32):7484–91.
72. Gartland PO. *Phys Norv* 1972;**6**:201.
73. Kahn A, Koch N, Gao W. Electronic structure and electrical properties of interfaces between metals and π-conjugated molecular films. *J Polym Sci B Polym Phys* 2003;**41**(21):2529–48.
74. Wandelt K. The local work function: concept and implications. *Appl Surf Sci* 1997;**111**:1–10.
75. Gadzuk JW. Inelastic resonance scattering, tunneling, and desorption. *Phys Rev B* 1991;**44**(24):13466.
76. Wingreen NS, Jacobsen KW, Wilkins JW. Inelastic scattering in resonant tunneling. *Phys Rev B* 1989;**40**(17):11834.
77. Dujardin G, Walkup RE, Avouris P. Dissociation of individual molecules with electrons from the tip of a scanning tunneling microscope. *Science* 1992;**255**(5049):1232–5.
78. Hla SW, Bartels L, Meyer G, Rieder KH. Inducing all steps of a chemical reaction with the scanning tunneling microscope tip: towards single molecule engineering. *Phys Rev Lett* 2000;**85**(13):2777–80.
79. Pascual JI, Lorente N, Song Z, Conrad H, Rust H-P. Selectivity in vibrationally mediated single-molecule chemistry. *Nature* 2003;**423**(6939):525–8.
80. Stipe BC, Rezaei MA, Ho W, Gao S, Persson M, Lundqvist BI, et al. Single-molecule dissociation by tunneling electrons. *Phys Rev Lett* 1997;**78**(23):4410–3.

81. Shin HJ, Jung J, Motobayashi K, Yanagisawa S, Morikawa Y, Kim Y, et al. State-selective dissociation of a single water molecule on an ultrathin MgO film. *Nat Mater* 2010;**9**(5):442–7.
82. Donarini A, Begemann G, Grifoni M. All-electric spin control in interference single electron transistors. *Nano Lett* 2009;**9**(8):2897–902.
83. Lee HJ, Ho W. Single-bond formation and characterization with a scanning tunneling microscope. *Science* 1999;**286**(5445):1719–22.
84. Pawin G, Wong K, Kim D, Sun D, Bartels L, Hong S, et al. A surface coordination network based on substrate-derived metal adatoms with local charge excess. *Angew Chem Int Ed* 2008;**47**(44):8442–5.
85. Schlickum U, Decker R, Klappenberger F, Zoppellaro G, Klyatskaya S, Ruben M, et al. Metal organic honeycomb nanomeshes with tunable cavity size. *Nano Lett* 2007;**7**(12):3813–7.
86. Stepanow S, Lin N, Barth JV. Modular assembly of low-dimensional coordination architectures on metal surfaces. *J Phys Condens Matter* 2008;**20**(18):184002.
87. Liljeroth P, Swart I, Paavilainen S, Repp J, Meyer G. Single-molecule synthesis and characterization of metal-ligand complexes by low-temperature STM. *Nano Lett* 2010;**10**(7):2475–9.
88. Nazin GV, Qiu XH, Ho W. Visualization and spectroscopy of a metal-molecule-metal bridge. *Science* 2003;**302**(5642):77–81.
89. Popov A, Shumkin G. A multiscale molecular switch model. *Moscow Univ Comput Math Cybern* 2010;**34**(3):106–12.
90. Sarhan Jr. A, Nelson BA, David M, Nakanishi H, Kasai H. STM-induced switching of the hydrogen molecule in naphthalocyanine. *J Phys Condens Matter* 2009;**21**(6):064201.
91. Blum AS, Kushmerick JG, Long DP, Patterson CH, Yang JC, Henderson JC, et al. Molecularly inherent voltage-controlled conductance switching. *Nat Mater* 2005;**4**(2):167–72.
92. Collier CP, Mattersteig G, Wong EW, Luo Y, Beverly K, Sampaio J, et al. A [2]Catenane-based solid state electronically reconfigurable switch. *Science* 2000;**289**(5482):1172–5.
93. Tao NJ. Electron transport in molecular junctions. *Nat Nanotechnol* 2006;**1**(3):173–81.
94. van der Molen SJ, Liljeroth P. Charge transport through molecular switches. *J Phys Condens Matter* 2010;**22**(13):133001.
95. Gross L, Mohn F, Liljeroth P, Repp J, Giessibl FJ, Meyer G. Measuring the charge state of an adatom with noncontact atomic force microscopy. *Science* 2009;**324**(5933):1428–31.
96. Gross L, Mohn F, Moll N, Meyer G, Ebel R, Abdel-Mageed WM, et al. The chemical structure of a molecule resolved by atomic force microscopy. *Science* 2009;**325**(5944):1110–4.
97. Joachim C, Gimzewski JK, Aviram A. Electronics using hybrid-molecular and mono-molecular devices. *Nature* 2000;**408**(6812):541–8.

Chapter 3

Electron Transfer Phenomena at the Molecular Scale: Organic Charge Transfer Complexes on Metal Surfaces

Isabel Fernández Torrente, Katharina J. Franke and Jose Ignacio Pascual
Institute für Experimentalphysik, Freie Universität Berlin, Arnimallee, Berlin, Germany

1. INTRODUCTION

Organic charge-transfer (CT) complexes are molecular compounds mixing two species with different electron affinities (EAs): an electron donor (D) and an electron acceptor (A). The donor molecule has a small ionization energy (IP), while the counterpart acceptor molecule has a large electronegativity or EA. The large difference in their chemical potentials leads to strong electron redistribution when they are brought into a mixture. The donor species oxidizes by the loss of charge and the acceptor is reduced, as described by the following reaction[1,2]:

$$[D_m]+[A_n] \rightarrow [D_m]^{\delta+}+[A_n]^{\delta-} \tag{3.1}$$

where δ is the CT ratio. The result is a molecular CT salt D_mA_n with specific new properties. CT compounds are investigated since decades but became of increasing importance recently in fields like, for example, molecular electronics or organic photovoltaics. The first suggestions of a molecular device, presented in 1974 by A. Aviram and M.A. Ratner, consisted in the combination of an electron donor species (TTF: tetrathiafulvalene, $C_6H_4S_4$) and an electron acceptor (TCNQ: 7,7,8,8-tetracyanoquidimethane, $C_{12}H_4N_4$).[3] More than 35 years have elapsed since this first prediction that set the basis of what today is known as 'molecular electronics'. There are many properties that make these compounds attractive for the creation of molecular devices: they are easy to grow, have a low cost, and specially, their properties can be tailored by means of organic synthesis.

Frontiers of Nanoscience, Vol. 2. DOI: 10.1016/B978-0-08-096355-6.00003-9

In a CT molecular solid, CT processes lead to an excess of charge carriers along certain directions, whose mobility depends on the crystalline structure and the nature of the molecular orbital overlapping.[4] Usually, the charge transferred between the molecular donor and acceptor has a dominantly π-electron character, resulting in the formation of delocalized electron bands. In some cases, these donor–acceptor interactions mediate the formation of CT crystalline solids in which the molecules are stacked in homomolecular rows. Within the rows, the molecular interactions are dominated by π–π overlap. The molecules pack rather densely in order to maximize the orbital overlapping between neighbouring molecules and, therefore, delocalizing the carriers throughout the crystal.

This is the particular case of bulk TTF–TCNQ, considered as a prototypical low-dimensional organic metal.[2] This CT salt is formed by the donor TTF and the acceptor TCNQ with a 1:1 stoichiometry (Figure 3.1A). It exhibits a

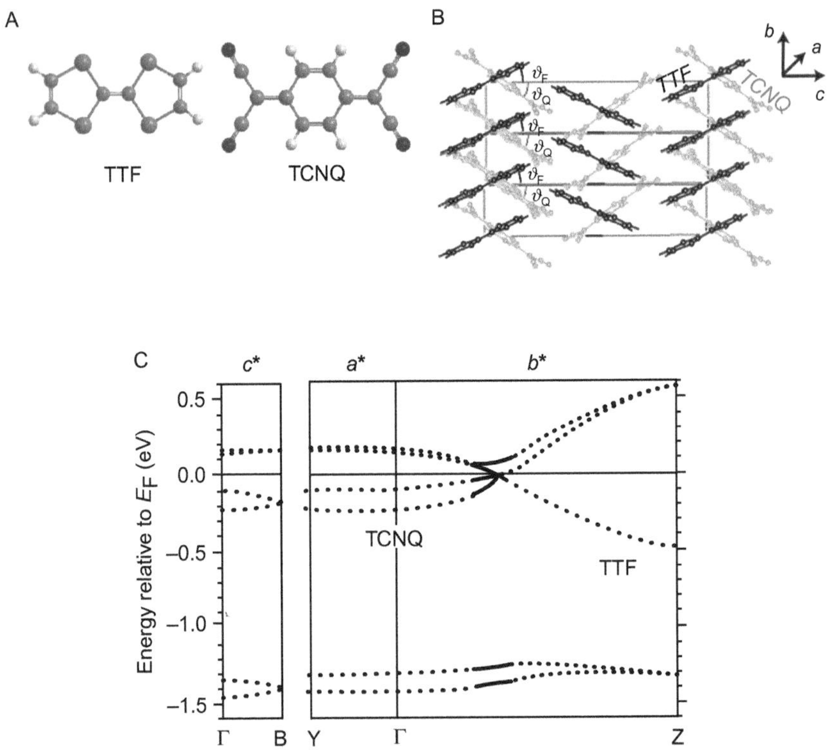

FIGURE 3.1 (A) Scheme of the TTF (left) and the TCNQ (right) molecules. Grey spheres stand for C atoms, white for H atoms, yellow for S atoms, and blue for N. (B) Bulk structure of TTF–TCNQ showing the principle directions, *a*, *b*, and *c*, of the three-dimensional unit cell. (C) Calculated band structure.[5] Dispersive behaviour is observed along the *b** direction. The splitting of the bands indicate that molecular chains are not completely isolated from each other. (For interpretation of the references to colour in this figure legend, the reader is referred to the Web version of this chapter.)

metal-like room temperature conductivity of 400 100 Ω^{-1} cm^{-1}.[6–8] TTF–TCNQ crystallizes in a monoclinic structure built up from homologous stacks of TTF and TCNQ. The molecules overlap within the rows in a 'ring double bond' fashion along direction *b* (Figure 3.1B). The metallicity of the compound is due to (i) the formation of bands through the π overlapping of the molecular components and (ii) the partial occupation of these bands at the Fermi level. The molecular packing maximizes the overlapping integral and, consequently, the amount of charge that can be delocalized along the stacks. The free carriers are generated in both π-stacked rows due to CT between TTF and TCNQ. This transferred charge is delocalized along the *b* direction of the crystal resulting in a quasi one-dimensional electron dispersion as it has been both described by theoretical DFT calculations[5,9,10] (Figure 3.1C) and observed experimentally.[11]

As a quasi one-dimensional molecular metal, TTF–TCNQ undergoes structural transformations associated to Peierls instabilities at low temperatures. Below 54 K, TTF–TCNQ experiences a sequence of phase transitions that progressively destroy the electronic transport[6] and affect the magnetic properties of the crystal.[12,13] There is a transitional region between $T = 54$ K and 38 K dominated by one-dimensional distortions[14,15] that develop charge density waves (CDW) on the *ab* surface of the crystal.[16,17]

In contrast with the existing exhaustive study of the bulk properties of TTF–TCNQ, very little is known about the thin-film behaviour, which have been studied only down to, approximately, 1 μm thickness on alkali halide substrates.[18] The transition from bulk TTF–TCNQ to ultra-thin films of monolayer thickness deposited on metals is expected to introduce new phenomenology related to the organic–inorganic interface. Effects like hybridization, CT with the surface and molecular level alignment become factors that may govern the electronic transport. Hence, the adsorption of an ultra-thin TTF–TCNQ layer on a metal opens a new field of research for the potential application of CT complexes as devices in the nanoscale.

The study of the organic–inorganic interface between a TTF–TCNQ layer and a metal surface is the aim of this chapter. Using low temperature scanning tunneling microscopy (STM) and spectroscopy (STS), our main goal is to resolve phenomena associated with charge redistribution of electron donor and acceptor species at the molecular scale. In the first sections of this chapter, the adsorption and charge configurations of individual TTF and TCNQ on Au(111) will be analysed. The mixed adsorption of TTF and TCNQ is also addressed. Even at the monolayer level, the donor–acceptor recognition is active, leading to the spontaneous formation of alternating bulk-like donor–acceptor rows. They exhibit though a very important difference with the bulk crystal: the molecules are not π-stacked with each other, but they adsorb planar on the surface. This difference, and the proximity to an infinite electron and hole reservoir, as it is the metal surface, leads to important differences in terms of the amount of CT at the limit of molecular monolayers.

2. TCNQ ON AU(111): THE NEUTRAL ADSORPTION OF AN ELECTRON ACCEPTOR MOLECULE

In spite of its strong acceptor character, TCNQ behaves as a rather weakly bound system when it is adsorbed on a Au(111) surface. TCNQ molecules appear in STM images at positive sample bias voltages with a characteristic intramolecular pattern composed of two U-like shapes at both sides of a nodal plane located at the middle of the quinonoid ring and an additional protrusion at the centre of the dicyanomethylene groups (Figure 3.2B and C). This shape strongly resembles the isosurface of lowest unoccupied molecular orbital (LUMO) resonance of the free molecule (Figure 3.2A). For negative bias voltages, TCNQ is imaged as a bright protrusion, without any internal molecular resolution (Figure 3.2D). The strong resemblance of the molecular orbitals to the LUMO isosurface of the free molecule supports the assumption of a weak physisorption. These results contrast to the adsorption of TCNQ and

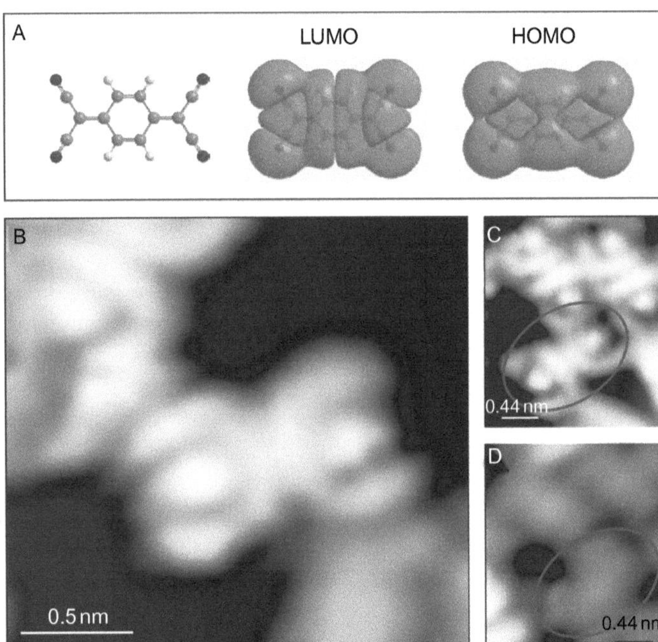

FIGURE 3.2 (A) Molecular model of TCNQ. Grey circles represent C, white H, and blue N. The grey surfaces are the isosurfaces corresponding to LUMO and HOMO of the relaxed free molecule. (B) TCNQ molecule imaged at positive bias voltage (V=0.8 V, I=0.32 nA). The LDOS resembles the LUMO shape of the free molecule, as can be observed by direct comparison with the isosurfaces presented in (A). (C) and (D) STM images of the same TCNQ group taken at positive (c: V=1 V, I=0.33 nA) and negative (d: V=−1 V, I=0.33 nA) bias voltages. At negative bias voltages, TCNQ is imaged without internal resolution. (For interpretation of the references to colour in this figure legend, the reader is referred to the Web version of this chapter.)

related molecules on copper surfaces, where chemical bonding of the CN terminations with copper atoms leads to a bent adsorption geometry and, in some cases, significant CT.[19–22]

Despite this weakly bounded physisorbed state, TCNQ exhibits a certain degree of commensuration with the Au(111) lattice. Its adsorption on a Au (111) surface results in the formation of a highly ordered molecular layer, that extends for hundreds of nanometers over the surface. The STM images show, superimposed to the molecular corrugation, the unperturbed pattern of the herringbone reconstruction of the Au(111) surface (Figure 3.3A), supporting a weak metal–organic interaction. The molecular structure formed has a rhombic unit cell with vectors $a_1 \approx a_2 \approx 1$ nm (Figure 3.3B). The vector a_1 runs along the [11–2] surface direction and a_2 presents a slight deviation of approximately 5° with respect to the [1–10] direction. As a result, TCNQ molecules alternate two adsorption sites, one with the C atoms of the central ring on Au(111) hollow sites and other where C atoms are distributed on bridge and top sites.

The self-assembled structure is stabilized via a saturated C. . .HC hydrogen bond network (Figure 3.3C). All N and H terminal atoms are directly involved in the formation of bonds. The interatomic distance amounts to 3 Å. This length is slightly larger than the interatomic hydrogen bond distances reported in liquid and gas phase complexes.[23] Therefore, even though TCNQ exhibits a weak adsorption on the metal surface, the formation of the CN. . .HC

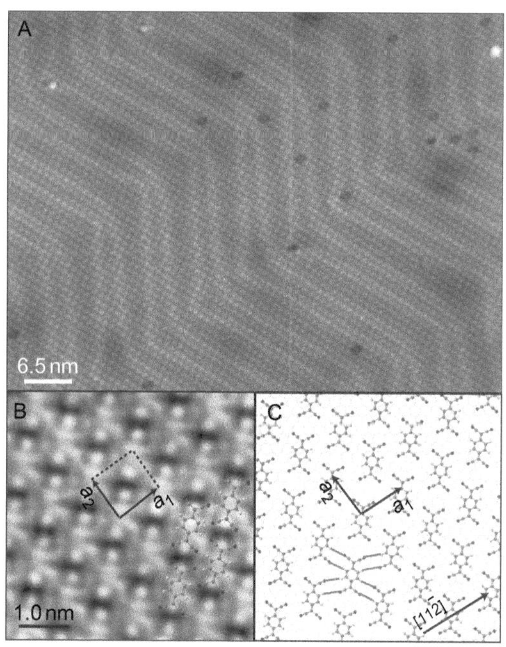

FIGURE 3.3 (A) Large STM area of a self-assembled domain of TCNQ. The reconstructed Au(111) surface is unaltered after molecular sublimation, as stated by the unperturbed herringbone reconstruction. (B) Zoom in a TCNQ island. The unit cell of the self-organized domain exhibits a rhombic symmetry. Four TCNQ molecular models are superimposed on the molecular lattice to clarify the orientation of the TCNQ monomers. (C) Adsorption model of TCNQ on Au(111). Molecules alternate two adsorption sites. Green lines mark the CN. . .HC bonds existing per molecule. All the H and N terminal atoms are involved in the attractive interaction. (For interpretation of the references to colour in this figure legend, the reader is referred to the Web version of this chapter.)

hydrogen bond network maintains a commensuration with the surface. The formation of such ordered structure is not in agreement with a tendency of the molecule to accept charge from the substrate. Despite the strong acceptor character of the molecule, there is no sign of CT between the molecule and the surface. We expect that either TCNQ is essentially neutral on the Au (111) surface, or the attractive nature of CN...HC bonds destroys the effect of electrostatic repulsion between the negatively charged molecules.

The differential conductance spectra on top of TCNQ molecules exhibit two characteristic fingerprints with respect to the bare Au(111) surface (Figure 3.4A). The first is an unoccupied resonance located at 0.7 eV.

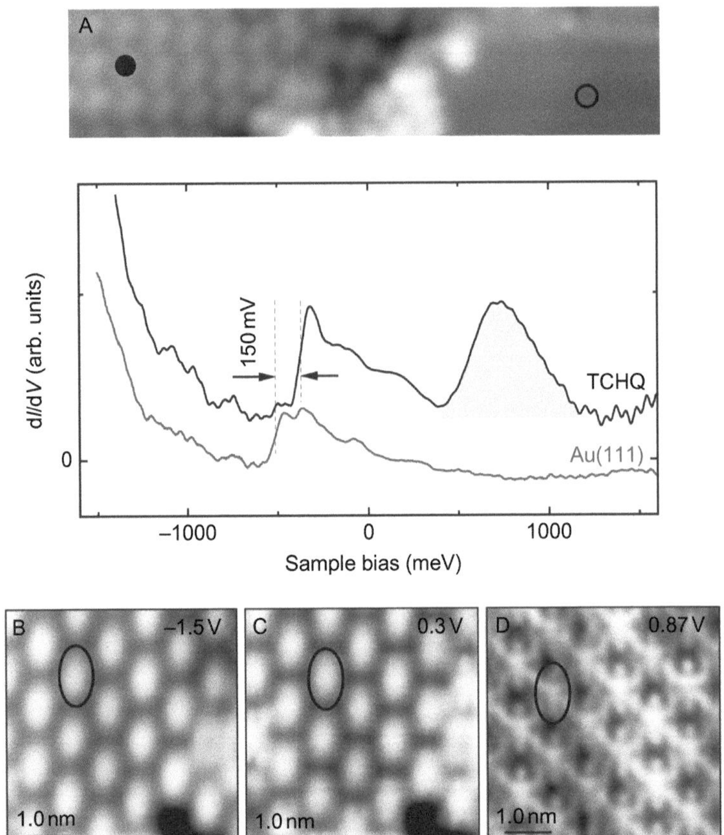

FIGURE 3.4 (A) STS spectra taken on clean Au(111) (red curve) and TCNQ island (black curve). Parameters: $V=2$ V, $I=1.4$ nA, $V_{ac}=7$ mV_{rms}. The black curve exhibits two features not observed on the conductance of the metal, namely a shift of the Au(111) surface state and an unoccupied resonance at 0.7 eV. (B)–(C) Topography images of a self-assembled TCNQ domain, taken at different voltages. The intramolecular TCNQ LUMO-shape resolution is obtained for bias voltages above 0.7 eV. (For interpretation of the references to colour in this figure legend, the reader is referred to the Web version of this chapter.)

The second is a shift of the Au(111) surface state of 150 meV towards the Fermi level. These two quantities are helpful to extract some information about the TCNQ/Au(111) interface. The molecular resonance can be easily identified with the TCNQ LUMO-derived orbital. STM images around this bias value show the clear features presented in Figure 3.2, whereas at different bias values, they appear rather shapeless.

The second spectroscopic fingerprint, an upward shift of the Au(111) surface state, has been also reported for other adsorbates on noble metal surfaces, like noble gases,[24–26] insulating thin films[27] and molecular layers.[1] Its origin is attributed to either modifications of the image potential and/or surface work functions by the dielectric medium placed above the metal surface[27–29] or to the surface state depopulation[30] produced by CT processes across a molecule/metal interface. Assuming that the upward shift of the surface state is due to electron depopulation, it is possible to estimate the amount of charge the surface would donate to the TCNQ molecule upon adsorption. The density of states of a (two-dimensional and free-electron like) surface state has a step-like shape with an onset at the binding energy (490 meV for clean Au(111)[31]) and can be approximated by the constant value $m^*\pi/h^2$, where m^* is the effective mass of the electrons in the Au (111) surface state, $m^* = 0.28\, m_e$.[32,33] Using this value, a linear shift amounting 150 meV of the surface state's onset corresponds to an average depopulation in ~0.2 e^- per molecular unit cell. This is a large value that would lead to considerable intermolecular electrostatic repulsion, on one hand, and to partial population of the LUMO resonance. Our STS measurements establish the LUMO position at 0.7 eV (Figure 3.4A), far away from E_F. Hence, the most probable origin of such shift of the Au(111) surface state is changed in the work function and/or image potential shape induced by the molecular layer.

3. TTF ON AU(111): FORMATION OF A WIGNER MOLECULAR LATTICE

The rather non-interacting behaviour of TCNQ molecules on Au(111) contrast with the chemically active behaviour of TTF on this metal surface. The existence of strong charge redistribution at the metal–organic interface can be directly inferred from STM images by analysing an intriguing effect in the distribution of TTF molecules on the surface. If TTF molecules are deposited on a surface at the temperature of 80 K (and, of course, cooled down to 5 K for STM imaging), TTF molecules appear randomly spread in the fcc and hcp regions of the Au(111) surface (Figure 3.5G). This disordered distribution is a characteristic of low-temperature adsorption. In this case, molecules are easily dragged by the STM tip, due to the weak adsorption on the surface. However, if a TTF pre-covered substrate is annealed to room temperature previous to its STM inspection, a characteristic ordered pattern appears. TTF align along the fcc

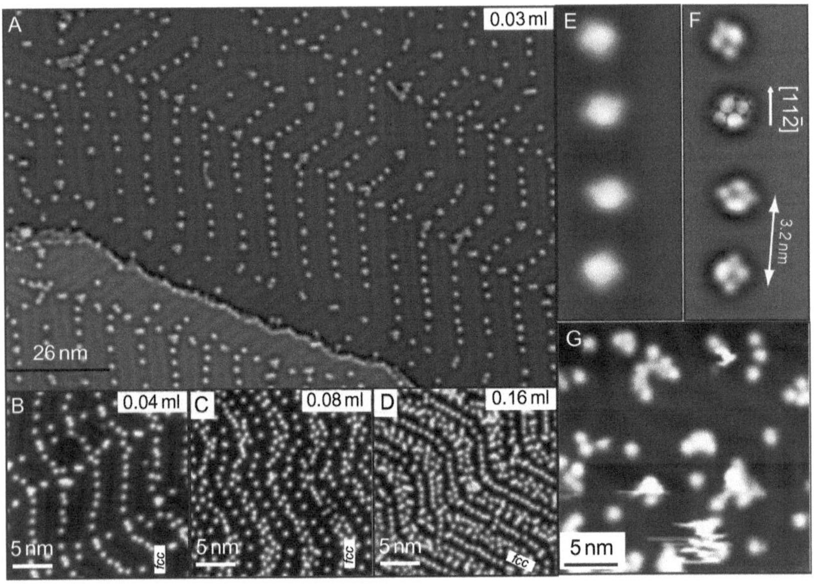

FIGURE 3.5 (A)–(D) STM images for different TTF coverages of 0.03 ML (V=0.9 V, I=0.3 nA) (A), 0.04 ML (V=1.2 V, I=0.2 nA) (B), 0.08 ML (V=0.8 V, I=0.1 nA) (C) and 0.16 ML (V=1.2 V, I=0.3 nA) (D). (E) STM image of four TTF monomers adsorbed on the fcc region of the reconstructed Au(111) surface (V=−1 V, I=0.3 nA), and (F) Laplace filtered image of (E). The STM signal is mainly located at the S atoms and the ethylene bonds as indicated by the TTF model superimposed to one TTF. The image reveals that two of the sulphur atoms for each molecule appear brighter suggesting a small tilt of the molecule plane with respect to the surface. (G) Deposition of molecules on a cold sample (80 K) leads to population of a weakly adsorbed precursor state, in which molecules may nucleate in clusters.

regions of the herringbone reconstruction in one-dimensional chains of TTF monomers separated by distances of about 2–3 nm, substantially larger than the length scale of intermolecular attractive interactions.

This behaviour prevails as the coverage is increased, accompanied by a monotonous decrease in the average pair distance between the monomers. We illustrate this in Figure 3.5A–D by representing the Au(111) surfaces with four different TTF coverages,[a] that is, 0.03, 0.04, 0.08 and 0.16 ML. For both 0.03 and 0.04 ML coverages, TTF only adsorbs in the fcc regions of the reconstructed Au(111) surface. Molecular adsorption at fcc sites might be favoured by the lower concentration of surface atoms compared to the hcp sites, where buried layers have to be taken into.account.[34] At a coverage of 0.08 ML, the array is compressed into a two-dimensional molecular distribution, and the hcp region starts to also be populated with similar one-dimensional-like arrays. As we shall show in this section, this peculiar superlattice is an expression of electrostatic

a. The molecular coverage is determined from STM images of large surface areas, assuming that 1 ML corresponds to two molecules per nm^2.

repulsion between 'charged' TTF monomers. Contrary to self-assembling of molecular structures via short-range non-covalent attractive interactions,[35] TTF molecules exhibit a clear tendency to maximize their intermolecular distance.

The intermolecular separation between molecular pairs follows a distribution characteristic of interacting one-dimensional systems, which depends on the specific coverage. A quantitative analysis of the one-dimensional average pair distance for the four different coverages investigated is shown in Figure 3.6. The statistics are performed over 500 molecular pairs per plot. In the cases of 0.03 and 0.04 ML coverages, the one-dimensional distribution statistics are carried out in the fcc domains, since hcp areas are not yet populated. On the other hand, for 0.08 and 0.16 ML, pair distances are measured only in hcp areas because of the two-dimensional distribution present in fcc sites at such

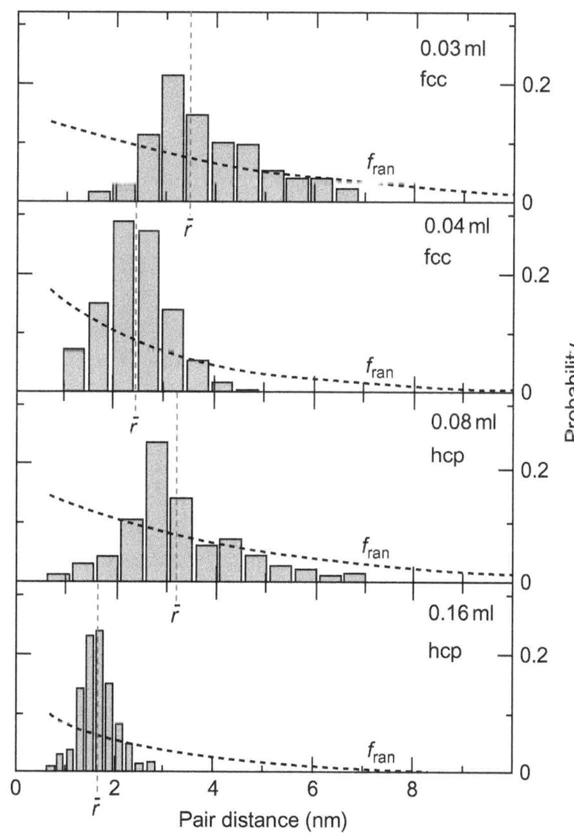

FIGURE 3.6 Pair distributions f of the one-dimensional TTF distributions for the data shown in Figure 3.5A–D. For 0.08 and 0.16 ML, the distributions are performed on hcp regions. The corresponding one-dimensional random distribution functions for non-interacting particles $f_{ran}(r)$ are included.

coverages. The distributions are peaked around an average pair distance of 3.5 nm (fcc), 2.5 nm (fcc), 3.3 nm (hcp) and 1.7 nm (hcp) for the lowest to the highest coverage, respectively.

A comparison with a one-dimensional random distribution $f_{ran}(r)$ of non-interacting particles for each coverage is plotted in Figure 3.6. The one-dimensional random separation distribution is expressed, in analogy to two-dimensional distributions, as[36]:

$$f_{ran}(r) = \frac{Nar(1 - 2r/L)^N}{L} \tag{3.2}$$

where N is the number of molecules per image, L is the length of the one-dimensional particle box and a is the discrete number of positions a molecule can occupy. In this limit of not interaction between particles, the nearest neighbour's pair distribution decays monotonously with the pair distance r. Therefore, the experimental peaked distributions obtained are symptomatic of a repulsive interaction. If we assume the system to be classical (distinguishable particles in a one-dimensional box) and in thermal equilibrium, experimental $f_{exp}(r)$ and random $f_{ran}(r)$ distributions can be related to each other by a Boltzmann factor.[37]

$$f_{exp}(r) = f_{ran}(r)\exp(-(\omega(r) - \mu)/k_B T) \tag{3.3}$$

where T is the temperature of the system at which the molecules were frozen in their observed position, k_B the Boltzmann's constant and $\omega(r)$ the mean intermolecular interaction potential energy. The chemical potential[b] term μ corresponds to the internal energy per molecule of a dense system of interacting particles and can be here approximated by the electrostatic energy per molecule in a fully periodic one-dimensional lattice.
From Equation (3.3), the following result can be deduced:

$$\frac{\omega(r)}{k_B T} = \frac{\mu}{k_B T} - \ln\left(\frac{f_{exp}}{f_{ran}}\right) \tag{3.4}$$

Equation (3.4) allows us to evaluate the factor $\omega(r)/k_B T$ as the combination of a zeroth order potential, $\mu/k_B T$, and a factor dependent on the deviation of the experimentally determined pair distribution from a random distribution. The results for the four different coverages of Figure 3.5A–D are presented in Figure 3.7.

b. This potential is defined, on a canonical collective described in terms of temperature, volume and number of particles, as the amount by which the energy of the system would change by the introduction (or removal) of a particle in the ensemble.[38] For the hypothetical case of infinite number of particles in a one-dimensional line the electrostatic interaction would be homogeneous among the uniformly distributed particles because there is no border effect. In that case the system is in equilibrium and it is possible to talk about a chemical potential.

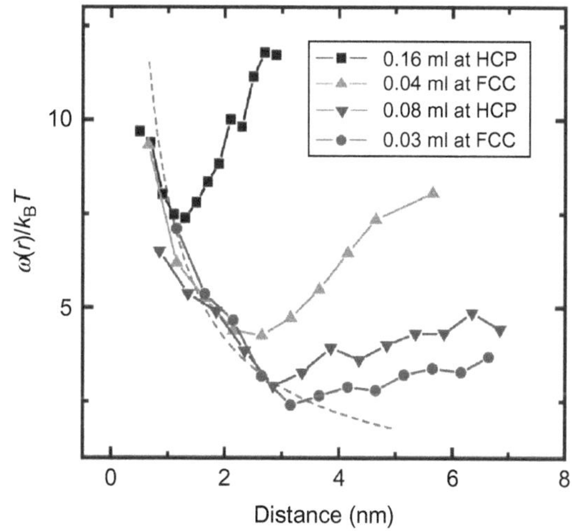

FIGURE 3.7 Mean interaction potentials $\omega(r)$ of one-dimensional TTF arrays obtained from the pair distributions shown in Figure 3.6. The dashed line represents the pair electrostatic interaction $E(r)$ between particles charged with 0.3 e^- and a temperature ($T = 165$ K) to fit the repulsive part of $\omega(r)$ for the most dilute case. Each curve has been shifted upwards an amount (8.4, 5.4, 4.1, 3.8, from top to bottom) representing the coverage dependent zeroth order internal potential, approximated here as the electrostatic energy per molecule charged with 0.3 electrons in a fully periodic lattice. The range of the interaction energy obtained from this fitting (55–114 meV) is considerably larger than the energy of a long-range interaction mediated by surface electrons.[36,39] Reprinted figure with permission from I. Fernandez-Torrente et al. Phys. Rev. Lett. 99, 176103, 2007. Copyright (2007) by the American Physical Society.

In the limit of a very dilute system, $\omega(r)$ is a good approximation for the repulsive pair interaction potential. At close pair distances, $\omega(r)$ decays as $1/r$, turning to constant value (hence converging with the random pair distribution) for large intermolecular separations. For higher coverages, interaction to both left and right neighbours of each monomer is important, and $\omega(r)$ has the shape of a potential well, more symmetric and shallower as we increase the density of TTF molecules. Such a shape arises from a sharper pair distribution and is consistent with the formation of a superlattice of TTF monomers.

An interesting conclusion from the experimentally obtained interaction potential is that the $1/r$ fall of $\omega(r)$ is consistent with repulsion between charged molecular monopoles, in contrast with the generally accepted idea that a local charge in the proximity of a metal behaves as a dipole, formed in addition to its image charge. Our results exclude that dipolar interactions between TTF molecules play a role. These are expected to decay much faster ($1/r^3$),[40] being negligible for molecules separated 2 or 3 nm. Thus, the monopolar shape of the interaction potential reveals that charges localized at the TTF monomers are not effectively screened by the underlying surface. This fact is surprising but can be interpreted considering that the Fermi wavelength of the Au(111) surface state is 3.9 nm and, hence, monomers at closer distances interact as underscreened particles.

From the results in Figure 3.7, we obtain that the interaction of a TTF monomer with its neighbours[c] lies between 55 and 114 meV, corresponding to the lowest and the largest coverage, respectively. To determine these values, we need to set the energy scale determined by k_BT, that is, the temperature at which thermal diffusion becomes negligible and the system freezes with the observed monomer distribution. This temperature can be associated to the threshold required to start the formation of the monomer superlattice, about 165 K, and thus can be used to extract a quantitative estimation of an electrostatic repulsion strength between monopoles charged with an average of 0.3 electrons. The interaction mechanism suggested from the dependence of $\omega(r)/k_BT$ on the pair distance and the coverage can be understood as follows: for low TTF densities on the metal surface, the repulsive interaction is purely coulombic. The aggregation of more charged molecules activates the creation of the superlattice, with shallower potentials located at the TTF monomers. This system can be compared, thus, to a one-dimensional molecular Wigner crystal created by CT localized at the metal–organic interface.[41]

An alternative source of long-range interactions between atomic and molecular adsorbates is Friedel oscillations of the two-dimensional electronic gas associated with the surface state.[32,36,39,42–46] A key element in this case is the oscillatory character of the interactions associated to half the Fermi wavelength λ_F, which amounts to a value of 1.8 nm for the Au(111) surface.[32] The STM images presented in Figure 3.5, and its statistical analysis, show a larger average pair distance than $\lambda_F/2$ and a monotonous decrease for larger concentrations of TTF molecules along the rows. Interaction mediated by surface electrons is also weaker, amounting to only a few millielectron volts for atomic adsorbates,[36,39] hence probably being here obscured by the stronger electrostatic repulsion. We can, however, find an unusual tendency of two TTF molecules to lie at approximately 1.8 nm, which reflects in the plots of Figure 3.7 as an ubiquitous flattening of the curves at this distance, as resembling a weaker potential minimum.

The origin of TTF behaving as an electric monopole can be rationalized by means of *ab initio* calculations. Density functional theory (DFT) simulations of TTF molecules on the unreconstructed Au(111) surface reveal an important charge redistribution driven by hybrid bonds between two sulphur atoms of the TTF molecule with Au atoms. This interaction leads to a tilted structure on the surface (Figure 3.8B and C) in agreement with an intrinsic asymmetry observed in the intermolecular structure in our STM images (Figure 3.5). At negative bias voltage, the STM image is basically dominated by the shape of the HOMO, with high density of states mainly at the S atoms and ethylene bonds, thus reflecting the dative character of the chemisorption.

---

c. The neighbours considered are located at distances r and $2r$.

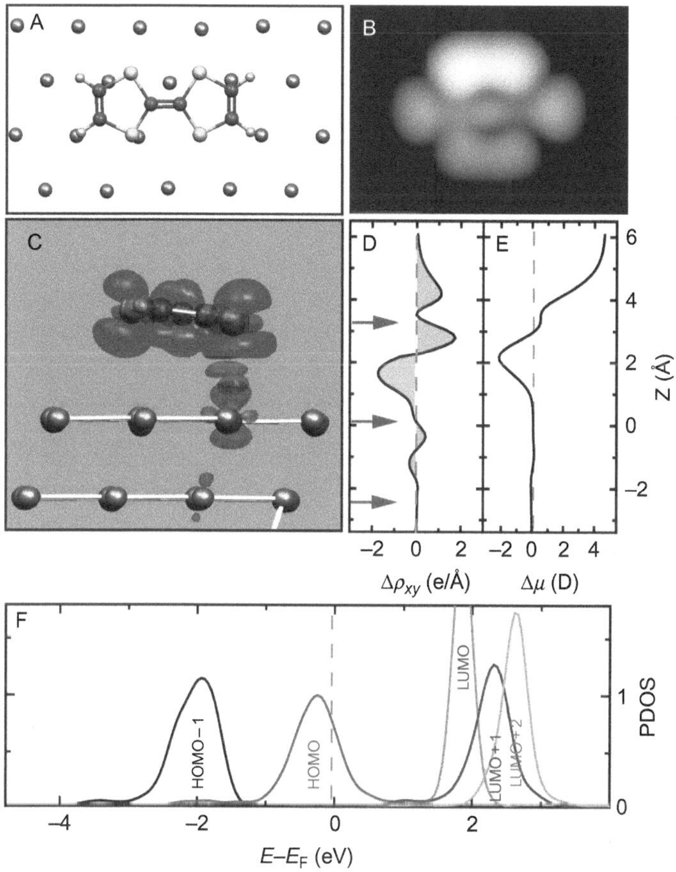

FIGURE 3.8 Results from DFT simulations. (A) Fully relaxed configuration of TTF on Au (111). (B) Tersoff–Hamman constant current image[47] of the molecule in (A) ($V=-0.5$ V). LDOS signal is mainly located on S atoms and on ethylene bonds. (C) Induced electronic density by the molecule–surface interaction. (D) Lateral (x–y planes) integration of the induced charge. The arrows show the vertical distance values at which the two topmost surface layers and the two binding S atoms lie. (E) Accumulated induced dipole. Together with (D) it reveals that the molecule becomes positively charged. (F) Projected density of states on molecular orbitals. The electronic states with HOMO character are partially empty, in agreement with the data of (C)–(E). Reprinted figure with permission from I. Fernandez-Torrente et al. Phys. Rev. Lett. 99, 176103, 2007. Copyright (2007) by the American Physical Society.

The analysis of the DFT results provides a quantitative estimation of the charge redistribution. As a consequence of chemical interaction, the HOMO becomes partially empty (Figure 3.8F), resulting in the positive charging of the molecule and the creation of a surplus of negative charge localized close to the S–Au bonds, as shown in the induced electronic density plot (Figure 3.8C). An excess of positive charge ($\approx$0.6 e^-) is located above the molecule, and the corresponding screening negative charge (≈-0.4 e^-) is between the molecule and the first atomic layer (Figure 3.8D), building up a dipole of 5.0 D (Figure 3.8E). The values of charge and dipole obtained from DFT simulations thus lead to similar conclusions as in our experiments: the electrical dipole built up at the interface is not large enough to govern interactions at pair separations above a few Åmstrongs.

3.1. TTF Nucleation: Long-Range Repulsive Versus Short-Range Attractive Interactions

While the adsorption of small coverages (<0.16 ML) leads to the formation of a superlattice characterized by the repulsive interaction between charged molecules, further addition of TTF on Au(111) results in molecular organization in self-assembled structures.

Upon deposition of approximately 0.5 ML, TTF forms zigzag chain-like structures uniformly distributed over the surface (Figure 3.9A and B). These

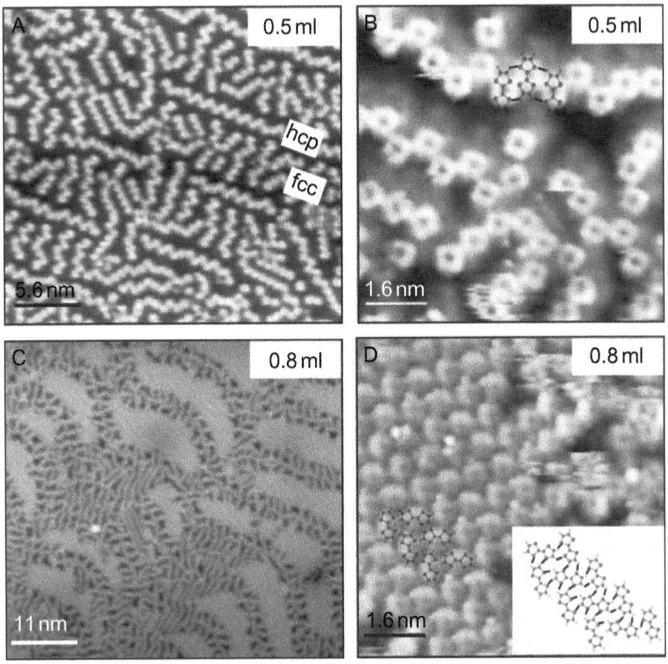

FIGURE 3.9 (A) Large STM image corresponding to a 0.5-ML TTF coverage ($V=-0.05$ V, $I=0.2$ nA). The molecules adsorb forming labyrinthine patterns dependent on the reconstructed Au(111) surface. (B) STM zoom in a small domain with sub-molecular resolution ($V=-0.03$ V, $I=0.34$ nA). The STM signal at the ethylene bonds allows to model the self-assembled structure via SHC bonds. (C) Large image corresponding to a 0.8-ML TTF coverage ($V=0.7$ V, $I=0.13$ nA). The molecules self-assemble in two-dimensional islands with a packed herringbone structure (D) ($V=-0.04$ V, $I=2.2$ nA). The inset shows a proposal for the saturated S...HC network that forms the islands.

molecular chains exhibit a labyrinthine pattern that varies on the hcp and fcc sites of the reconstructed Au(111) surface. While in the fcc domains the chains tend to have a short length and several orientations, in the hcp regions the tendency is reversed with a preferential formation of long chains along the soliton lines. STM images with intramolecular resolution resolve the relative orientation of the TTF molecules within the zigzag chains (Figure 3.9B). TTF molecules adsorb slightly shifted within the chains, being the structure dominated by S...HC hydrogen bonds.[48]

For coverage values approaching one monolayer, two-dimensional TTF islands start to appear (Figure 3.9C and D). Here, molecules pack densely in a parquet-like structure where adjacent molecules are rotated 60° to form a saturated S...HC network. The areas covered by one-dimensional TTF chains exhibit the same zigzag structure previously observed in the 0.5 ML coverage.

The characteristic labyrinthine pattern obtained for large TTF densities is a fingerprint of a system grown with competing long-range repulsive and short-range attractive interactions.[49–51] The strength of the competing interactions varies with TTF density. In the low coverage region, below ≈0.2 ML, the repulsive interaction is dominant and is the driving force responsible for the monomer superlattice formation. For medium coverages, ranging between ≈0.2–0.8 ML, both repulsive and attractive interactions play a role in the TTF self-organization resulting in a delicate balance. Above ≈0.8 ML island formation is predominant, as a consequence of the dominant short-range attractive interaction.

4. TTF–TCNQ ON AU(111): MOLECULAR MAGNETISM INDUCED BY CT

The co-adsorption of TTF and TCNQ on Au(111) leads to the spontaneous formation of a row structure alternating TTF and TCNQ (Figure 3.10A), revealing a donor–acceptor recognition pattern. The structure of the molecular monolayer differs from the characteristic π-stacks of the bulk crystal. In the single layer, molecules do not interact among them through the π-orbitals, but they adsorb planar on the surface. This planarity, together with the presence of the surface, leads to important differences in the charge reorganization at the interface. High-resolution tunnelling spectroscopy on TCNQ molecules finds a zero-bias peak (ZBP), shown in Figure 3.10B. The shape of the ZBP is close to a Lorentzian with a line width of approximately 8 ± 1 mV. STS performed in the neighbouring TTF molecules does not show any special fingerprint at that energy (Figure 3.10B); therefore, the signal corresponds unequivocally to the TCNQ molecule located in the self-assembled CT complex. The narrow width of the ZBP rules out the direct association of this peak to a molecular resonance. Instead, this sharp state pinned at zero bias is a clear fingerprint related to the existence of a magnetic moment in the proximity of an electron sea, that is, a Kondo resonance.

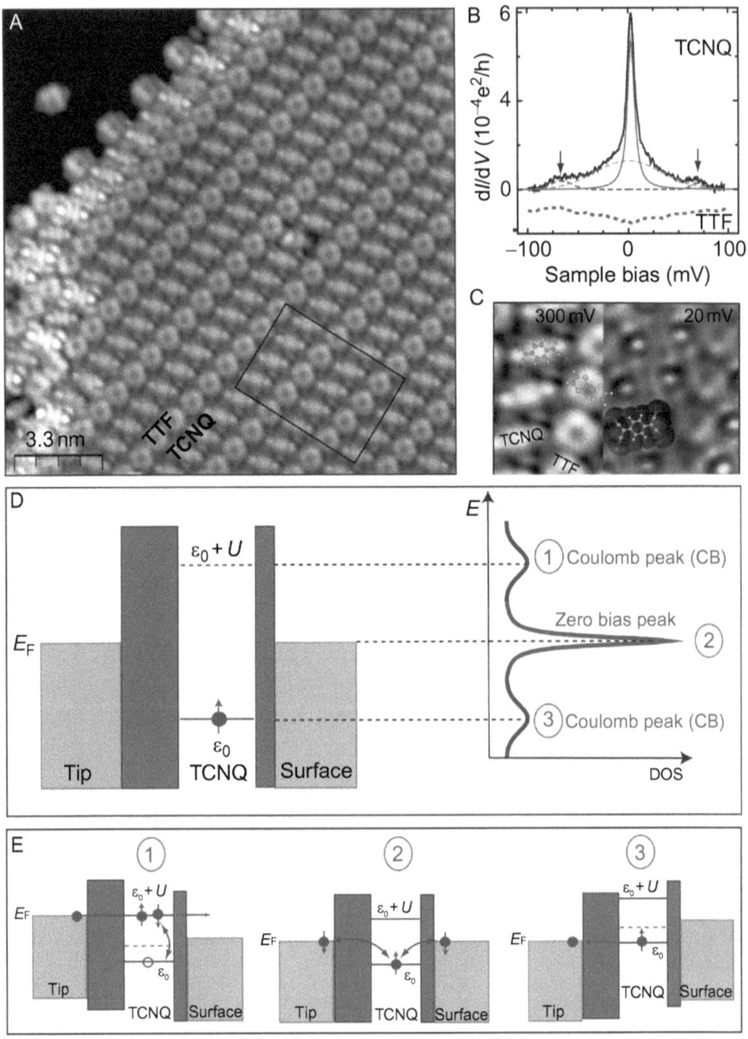

FIGURE 3.10 (A) STM images of a self-assembled TTF–TCNQ layer. (B) d*I*/d*V* spectra obtained in a TCNQ and in a TTF molecule ($I = 1$ nA, $V = 0.4$ V, Vac = 1 mVrms). The Gaussian fitting enhances the position of both the ZBP at the Fermi level (magenta line) and the satellite Coulomb peaks corresponding to singly and doubly occupied LUMO (marked with arrows). (C) Topography images at bias voltages of 300 and 20 mV. The LUMO orbital fingerprint at the cyano groups, missing in the TTF–TCNQ mixed phase for high voltages is recovered at voltages close to the Fermi level. The LUMO isosurface of the free TCNQ is also plotted for comparison. (D) Scheme of the density of states expected in the exchange process. The central peak at Fermi (ZBP) arises from the spin–flip mechanism and the Coulomb peaks (CB) correspond to the occupation of resonances ε_0 and ε_0 U. (E) Schemes of the tunnelling processes leading to the resonances of (A). 1 and 3 represent resonant tunnelling through ε_0 and $\varepsilon_0 + U$ CBs. 2 represents the spin–flip that occurs via virtual bound states at the Fermi level of the metal electron. (For interpretation of the references to colour in this figure legend, the reader is referred to the Web version of this chapter.)

The spin Kondo effect is a many-body problem that arises from the interaction between a single magnetic atom/molecule with an intrinsic spin and the electrons (spins) in a non-magnetic material. The result of this interaction is the creation of a bound state between the impurity spin and the spin of electrons in the non-magnetic material. At temperatures below a certain value known as Kondo temperature (T_K), the formation of a Kondo bound state provokes the arising of a new channel of conduction at the Fermi level of the system, observed as a many-body resonance at the Fermi energy in the density of states. The physics involved in the Kondo problem can be understood from the perspective of the Anderson impurity model, which was developed to explain the formation of magnetic moments in a metal.[52] The simplest Anderson model features the interaction of one local moment related to a magnetic impurity and the host electrons in the metal. As sketched in Figure 3.10D, the addition of an extra spin in the local state ε_0 costs an extra energy U, related to the Coulomb repulsion between the two charges. The doubly occupied level does have an energy $\varepsilon_0 + U$. For sufficiently low temperature, exchange processes that effectively flip the spin of the impurity can take place. This exchange is related to the creation of the Kondo bound state between the magnetic and the metal spins. The spin–flip processes occur via virtual intermediate states whose lifetime is low enough to provide the sufficient energy for the spin to jump between the local state and the Fermi sea as sketched in Figure 3.10D. This exchange process occurs at the Fermi level of the metal. As a consequence of the spin–flip process,[53] the density of states increases when the Fermi levels of tip and sample are aligned, forming the so-called Kondo–Abrikosov–Suhl resonance, the ZBP observed in the spectra. The extra Coulomb Blockade (CB) peaks appear by resonant tunnelling processes when either the Fermi level of the tip or the surface is aligned with the molecular levels ε_0 or $\varepsilon_0 + U$ (Figure 3.10E). The lineshape of the resonance at the Fermi level is described as a Fano lineshape[54,55] and varies between a Lorentzian peak and a dip, being the former considered as the pure Kondo process. Lineshapes different to a Lorentzian ZBP arise by interference effects between the Kondo channel and other tunnelling channels.

The Anderson model describes several possible regimes, depending on the values of three parameters: the energy of the magnetic impurity state (ε_0), the Coulomb repulsion (U) and the resonance width (Γ). Spin Kondo effect appears when ε_0 is singly occupied and localized (the CB peaks at ε_0 and $\varepsilon_0 + U$ do not overlap), but the resonance's line width Γ is finite (there is interaction with the metal). If, on the contrary, the interaction increases and Γ becomes comparable to U, the spin polarization of the level ε_0 vanishes because quantum spin fluctuations become possible in the impurity due to overlapping of both CB levels. This is a regime of mixed valence in which the magnetic character of the ground state vanishes. The lineshape of the ZBP also varies with the temperature. It broadens, reduces its intensity and vanishes for $T > T_K$. Above T_K, the CB peaks overlap and the spin polarization of the bound state disappears. As we shall show later, the temperature dependence of the line width and peak height of the ZBP can be used to decipher the Kondo temperature of the system.

In agreement with the Anderson model, the spectra on TCNQ molecules show two broader resonances (FWHM≈25 mV) located at ≈67 mV, symmetric with respect to the central peak (Figure 3.10B). Following the scheme of Figure 3.10D, the peak at −67 mV corresponds to tunnelling through the singly occupied energy level ε_0, whereas its symmetric resonance, at +67 mV, corresponds to tunnelling through the doubly occupied resonance level at ε_0+U. It is then possible to obtain a value of the intramolecular Coulomb repulsion U of 134 meV, much smaller than the expected value for the free molecule screened by the underlying metal surface. The pronounced Lorentzian shape further reveals the absence of interference effects with other tunnelling channels, as it is the case of Kondo resonances with different Fano lineshapes.

A further verification of the Kondo nature of the ZBP involves the study of its lineshape dependence on the temperature. Figure 3.11 plots the ZBP lineshape for several temperatures ranging from 4.8 to 13.5 K and the dependence of both the width (b) and the intensity (c) of the peak with the temperature. The line width broadens following an approximated expression developed in

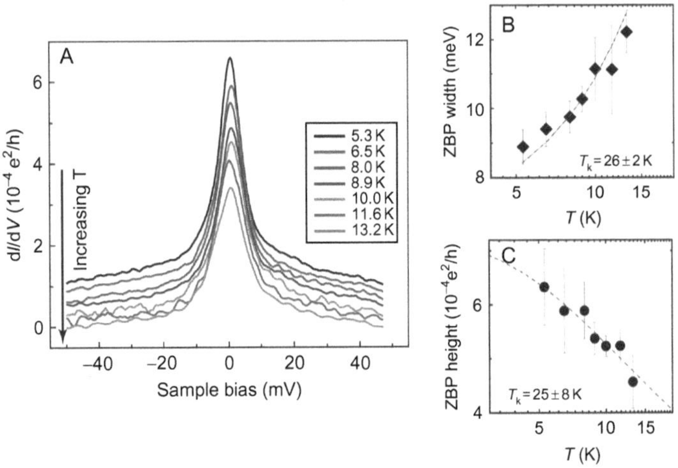

FIGURE 3.11 (A) Lineshape dependence of the zero-bias resonance on the temperature ($I=0.9$ nA, $V=100$ mV, $V_{ac}=0.3$ mV$_{rms}$). The resonance becomes broader and less intense for increasing temperatures. (B and C) Behaviour of the FWHM (B) and the zero-bias peak intensity (C) with the temperature. The points are an average over 6–16 measurements per each temperature. The red line fitting follows equations of Refs. 53 and 56. Reprinted figure with permission from I. Fernandez-Torrente et al. Phys. Rev. Lett. 101, 217203, 2008. Copyright (2008) by the American Physical Society.

the framework of the Fermi liquid theory,[53] resulting in a Kondo temperature of $T_K = 26 \pm 2$ K. In a similar way, the suppression of the peak height can be fitted to the empirical expression

$$G(T) = G_0 \left(1 + \frac{T^2 2^{1/S} - 1}{T_K^2}\right)^{-s} \quad (3.5)$$

which reproduces numerical renormalization group results for spin 1/2 systems[56] with the parameter $s = 0.22$ and with $T_K{}^* = T_K/(2^{1/s} - 1)^{1/2}$. In this case, the obtained Kondo temperature has a similar value of $T_K = 25 \pm 8$ K.

The Kondo effect is thus surprising in this non-metallic system and reveals the existence of an anionic ground state for TCNQ molecules within the TTF–TCNQ molecular layer. This state can be realized through the electron transfer of the TTF+Au(111) environment into a molecular resonance ε_0, here identified as the LUMO. The Kondo state further requires a certain degree of charge localization, which according to the Anderson model corresponds to $U > \Gamma$. Hence, the spectroscopic results fit with the canonical description of a Kondo process and leads to the logical implication of having an organic system with a certain degree of spin polarization. Interesting effects also appear from the fact of involving a molecular state conjugated into an extended carbon backbone, as one can realize by investigating the spatial distribution of the Kondo resonance.

The strength of the Kondo ZBP is not homogeneous over the molecule but exhibits a strong dependence on the intramolecular site where it is measured. Spectra as in Figure 3.10B are obtained at the cyano end groups. The peak intensity decreases gradually as we approach the centre of the TCNQ molecule (Figure 3.12A and B). Constant height conductance maps taken at energies close to zero bias resolve the spatial distribution of the Kondo peak intensity (Figure 3.12C) following the spatial shape of the LUMO resonance. Its intensity is maximal at the CN group and decreases on the centre of the molecule, in agreement with the spectroscopy taken locally. The decrease of ZBP intensity is accompanied by the arising of two extra side peaks at energies of $\approx$41 mV, with an intensity comparable to the Kondo resonance at the centre of the molecule (Figure 3.13A). An additional feature in the dI/dV spectra observed in TCNQ is the emergence of step-like functions for bias voltages larger than the side bands. The origin of both the lateral side peaks and step-like functions is related to the coupling of the unpaired electron with vibrations in the molecule as sketched in Figure 3.13B. These processes are possible because of the π-orbital nature of the orbital hosting the spin and its ubiquitous coupling to molecular vibrations.[57–62] Theoretical predictions describe the viability of an inelastic Kondo effect mediated by vibron-assisted tunnelling.[63,64] The charge tunnelling at energies close to the ZBP causes a slight distortion of the molecule and excites vibrational modes (vibronic coupling). The zero-bias resonance is not suppressed by the vibronic

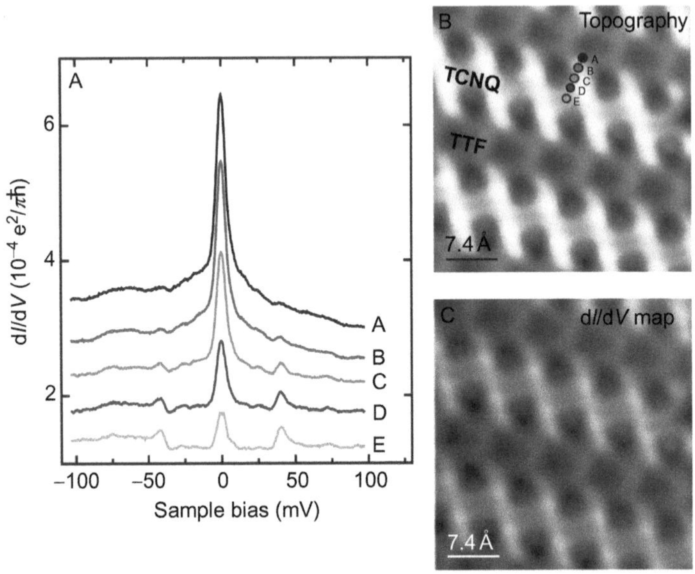

FIGURE 3.12 (A) Plots 5 spectra taken along a line scan from the CN group to the centre of a TCNQ molecule ($I=1$ nA, $V=100$ mV, $V_{ac}=0.7$ mV$_{rms}$). The direction of the scan is shown on the topography image (B). Panel (C) presents a conductance map taken in the constant height mode ($V=3$ mV, $I=0.1$ nA, $V_{ac}=2$ mV$_{rms}$). The Kondo signal has its maximum intensity at the CN groups and decreases gradually towards the centre of the TCNQ molecule.

coupling but breaks up into vibron side bands. These satellite peaks represent the resonant tunnelling events that involve both spin–flip plus excitation of a vibration. In our case, the symmetric position of the peaks at $\approx$41 meV (Figure 3.13A) with respect to the Fermi level and their narrow lineshape constitute a fingerprint of this strong electron–phonon coupling taking place at the centre of the TCNQ molecule.

The origin of the strong electron–phonon coupling can be understood by an examination of the atomic motion of the vibrational modes exhibited by the TCNQ at energies close to 40 meV. According to the literature,[65,66] this energy corresponds to a planar breathing mode, ν_9, with a_g symmetry that involves a distortion of the central ring and a stretching of the $CC(CN)_2$ double bonds of the $TCNQ^-$ anion. It is also associated to a shift of the LUMO position, thus driving charge out/into the molecule. In particular, the ν_9 mode found here exhibits a large electron–vibration coupling. It fits with the structural distortions undergone by the $TCNQ^-$ anion upon addition (removal) of charge, namely an increase (decrease) in the aromatization of the central ring and the decrease (increase) of the external CC bond order (Figure 3.13C).

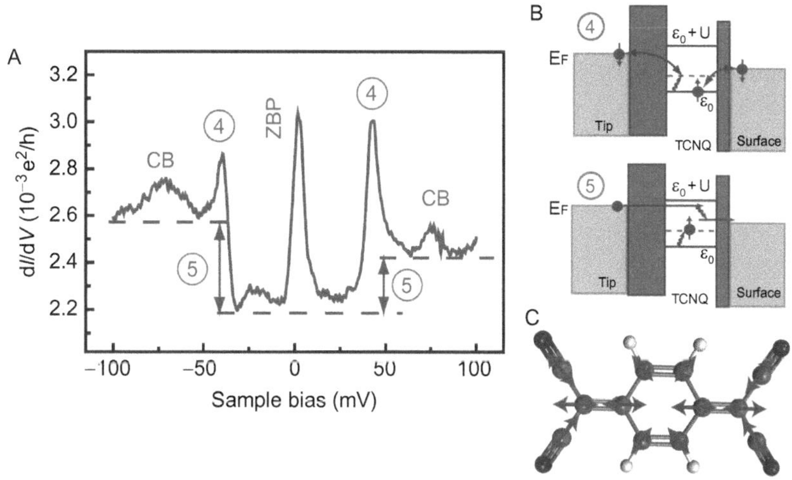

FIGURE 3.13 (A) dI/dV curve taken on the centre of a TCNQ molecule ($I = 1$ nA, $V = 100$ mV, $V_{ac} = 0.7$ mV$_{rms}$). Two vibronic sidebands (4) arise at symmetric values with respect to the Fermi level, $V = \pm 41$ mV. They involve both spin–flip and the excitation of a vibration. The dash line (5) sketches the characteristic step-like shape of a non-resonant electron transfer process. Panel (B) schematizes the elastic (4) and the inelastic (5) processes taking place at the molecular centre. (C) Structural changes, marked with arrows, of TCNQ upon charge injection at ± 41 mV. This vibrational mode is characterized by a strong distortion of the central quinonoid ring, which becomes fully aromatic for the doubly ionized TCNQ.

5. CONCLUSIONS AND REMARKS

In this chapter, we have investigated the adsorption of a model CT complex composed by the electron donor TTF and electron acceptor TCNQ on a Au (111) surface. Using a combination of local spectroscopy and microscopy, we have resolved diverse electron redistribution processes taking place at the organic–metal interface. First, we have analysed the adsorption of each component separately. The weak physisorption of TCNQ contrasts with the chemisorption of TTF, driven by covalent S–Au bonds. This results in a CT that positively charges the TTF molecules. As a consequence of this charge state, the TTF monomers form a lattice based on long-range repulsive interactions of electrostatic origin. The characteristic pattern of the Au(111) herringbone reconstruction imposes a one-dimensional character to the repulsive lattice. The system can be thus considered as a molecular Wigner crystal, where each molecule defines a spatial location at the interface with charge accumulation.

The most surprising consequence of the charge redistribution is observed for the mixed adsorption of both TTF and TCNQ. Molecular domains with 1:1 stoichiometry assemble according to their donor–acceptor character in alternating rows, similar to the bulk organization. In the self-assembled phase, TCNQ loses its neutrality and is charged with one electron, as we could detect with the observation of a Kondo resonance in these species. Through analysis of the temperature dependence of the lineshape of the Kondo peak, we could find the system lies in a spin 1/2 state, with a Kondo temperature of approximately 26 K. Further, due to the molecular nature of TCNQ, the Kondo peak is accompanied by fingerprints related to vibrational phenomena inherent in molecules. Vibronic-like side bands and inelastic steps are resolved in the conductance curves taken at the centre of the molecule reflecting the π-orbital character of the unpaired electrons.

In view of these, we can say organic donor–acceptor interactions are thus a promising avenue towards the spontaneous self-assembled organization of metal-free molecular magnets. The π-orbital character of the unpaired electron provides new phenomenology (vibrations) in comparison with the heavy atom-based magnetism. The study of the electronic and magnetic properties in nanostructures at the single molecule scale provides a good insight into the variables affecting the transport inside molecular complexes. This knowledge, together with chemical custom-made synthesis, may allow in the future an effective functionalization of molecular-based devices.

REFERENCES

1. Nicoara N, Román E, Gómez-Rodríguez JM, Martín-Gago JA, Méndez J. Scanning tunneling and photoemission spectroscopies at the PTCDA/Au (111) interface. *Org Electron* 2006;**7**:287–94.
2. Toyota N, Lang M, Muller J. *Low-dimensional molecular metals.* Springer; 2007.
3. Aviram A, Ratner MA. Molecular rectifiers. *Chem Phys Lett* 1974;**29**:277–83.
4. Fraxedas J, editor. *Molecular organic materials: from molecules to crystalline solids.* Cambridge University Press; 2006.
5. Sing M, Schwingenschlgl U, Claessen R, Blaha P, Carmelo JMP, Martelo LM, et al. Electronic structure of the quasi-one-dimensional organic conductor TTF-TCNQ. *Phys Rev B* 2003;**68**:125111.
6. Coleman LB, Cohen MJ, Sandman DJ, Yamagishi FG, Garito AF, Heeger AJ. Superconducting fluctuations and the Peierls instability in an organic solid. *Solid State Commun* 1973;**12**:1125–32.
7. Ferraris J, Cowan DO, Walatka V, Perlstein JH. Electron transfer in a new highly conducting donor-acceptor complex. *J Am Chem Soc* 1973;**95**:948–9.
8. Thomas GA, Schafer DE, Wudl F, Horn PM, Rimai D, Cook JW, et al. Electrical conductivity of tetrathiafulvalenium-tetracyanoquinodimethanide (TTF-TCNQ). *Phys Rev B* 1976;**13**:5105–10.
9. Fraxedas J, Lee YJ, Jimnez I, Gago R, Nieminen RM, Ordejón P, et al. Characterization of the unoccupied and partially occupied states of TTF-TCNQ by XANES and first-principles calculations. *Phys Rev B* 2003;**68**:19.
10. Ishibashi S, Kohyama M. Ab initio pseudopotential calculation for TTF-TCNQ and TSeF-TCNQ. *Phys Rev B* 2000;**62**:7839–44.
11. Zwick F, Jrome D, Margaritondo G, Onellion M, Voit J, Grioni M. Band mapping and quasiparticle suppression in the one-dimensional organic conductor TTF-TCNQ. *Phys Rev Lett* 1998;**81**:2974–7.

12. Gunning WJ, Chiang CK, Heeger AJ, Epstein AJ. Effects of controlled disorder on the phase-transitions and charge-density wave state in tetrathiofulvalene-tetracyanoquinodimethane. *Phys Status Solidi B* 1979;**96**:145–51.
13. Takahashi T, Jerome D, Masin F, Fabre JM, Giral L. 13C NMR studies of TTF (13C)-TCNQ. *J Phys C* 1984;**17**:3777–92.
14. Comés R, Shapiro SM, Shirane G, Garito AF, Heeger AJ. Neutron-scattering study of the 38- and 54-K phase transitions in deuterated tetrathiafulvalene-tetracyanoquinodimethane (TTF-TCNQ). *Phys Rev Lett* 1975;**35**:1518–21.
15. Denoyer F, Comés F, Garito AF, Heeger AJ. X-ray-diffuse-scattering evidence for a phase transition in tetrathiafulvalene tetracyanoquinodimethane (TTF-TCNQ). *Phys Rev Lett* 1975;**35**:445–9.
16. Nishiguchi T, Kageshima M, Kato NA, Kawazu A. Behavior of charge density waves in a one-dimensional organic conductor visualized by scanning tunneling microscopy. *Phys Rev Lett* 1998;**81**:3187–90.
17. Wang ZZ, Girard JC, Pasquier C, Jrome D, Bechgaard K. Scanning tunneling microscopy in TTF-TCNQ: phase and amplitude modulated charge density waves. *Phys Rev B* 2003;**67**:121401.
18. Fraxedas J, Molas S, Figueras A, Jiménez I, Gago R, Senzier PA, et al. Thin films of molecular metals TTF-TCNQ. *J Solid State Chem* 2002;**168**:384–9.
19. Bedwani S, Wegner D, Crommie MF, Rochefort A. Strongly reshaped organic-metal interfaces: tetracyanoethylene on Cu(100). *Phys Rev Lett* 2008;**101**:216105.
20. Choi T, Bedwani SP, Rochefort A, Chen CY, Epstein AJ, Gupta JA. A single molecule Kondo switch: multistability of tetracyanoethylene on Cu(111). *Nano Lett* 2010;**10**:4175–80.
21. Romaner L, Heimel G, Brdas JL, Gerlach A, Schreiber F, Johnson RL, et al. Impact of bidirectional charge transfer and molecular distortions on the electronic structure of a metal-organic interface. *Phys Rev Lett* 2007;**99**:256801.
22. Tseng T-C, Urban C, Wang Y, Otero R, Tait SL, Alcamí M, et al. Charge-transfer-induced structural rearrangements at both sides of organic/metal interfaces. *Nat Chem* 2010;**2**:374–9.
23. Hippler M. Quantum chemical study and infrared spectroscopy of hydrogen-bonded $CHCl_3$–NH_3 in the gas phase. *J Chem Phys* 2007;**127**:084306.
24. Andreev T, Barke I, Hövel H. Adsorbed rare-gas layers on Au(111): shift of the Shockley surface state studied with ultraviolet photoelectron spectroscopy and scanning tunneling spectroscopy. *Phys Rev B* 2004;**70**:205426.
25. Forster F, Nicolay G, Reinert F, Ehm D, Schmidt S, Höfner S. Surface and interface states on adsorbate covered noble metal surfaces. *Surf Sci* 2003;**532**:160–5.
26. Park JY, Ham UD, Kahng SJ, Kuk Y, Miyake K, Hata K, et al. Modification of surface-state dispersion upon Xe adsorption: a scanning tunneling microscope study. *Phys Rev B* 2000;**62**:R16341–R16344.
27. Repp J, Meyer G, Rieder KH. Snell's law for surface electrons: refraction of an electron gas imaged in real space. *Phys Rev Lett* 2004;**92**:036803.
28. Hotzel A, Moos G, Ishioka K, Wolf M, Ertl G. Femtosecond electron dynamics at adsorbate-metal interfaces and the dielectric continuum model. *Appl Phys B* 1999;**68**:615–22.
29. Smith NV. Phase analysis of image states and surface states associated with nearly-free-electron band gaps. *Phys Rev B* 1985;**32**:3549–55.
30. Morgenstern K, Braun KF, Rieder KH. Surface-state depopulation on small Ag(111) terraces. *Phys Rev Lett* 2002;**89**:226801.

31. Reinert F, Nicolay G, Schmidt S, Ehm D, Höfner S. Direct measurements of the L-gap surface states on the (111) face of noble metals by photoelectron spectroscopy. *Phys Rev B* 2001;**63**:115415.
32. Hyldgaard P, Persson M. Long-ranged adsorbate-adsorbate interactions mediated by a surface-state band. *J Phys Condens Matter* 2000;**12**:L13–L19.
33. Kevan SD, Gaylord RH. High-resolution photoemission study of the electronic structure of the noble-metal (111) surfaces. *Phys Rev B* 1987;**36**:5809–18.
34. Barth JV, Brune H, Ertl G, Behm RJ. Scanning tunneling microscopy observations on the reconstructed Au(111) surface: atomic structure, long-range superstructure, rotational domains, and surface defects. *Phys Rev B* 1990;**42**:9307–18.
35. Barth JV, Costantini G, Kern K. Engineering atomic and molecular nanostructures at surfaces. *Nature* 2005;**437**:671–9.
36. Knorr N, Brune H, Epple M, Hirstein A, Schneider MA, Kern K. Long-range adsorbate interactions mediated by a two-dimensional electron gas. *Phys Rev B* 2002;**65**:115420.
37. Huang K. *Statistical mechanics*. New York: Wiley; 1987.
38. Baierlein R. *Thermal physics*. Cambridge University Press; 1999.
39. Repp J, Moresco F, Meyer G, Rieder KH, Hyldgaard P, Persson M. Substrate mediated long-range oscillatory interaction between adatoms: Cu/Cu(111). *Phys Rev Lett* 2000;**85**:2981–4.
40. Yokoyama T, Takahashi T, Shinozaki K, Okamoto M. Quantitative analysis of long-range interactions between adsorbed dipolar molecules on Cu(111). *Phys Rev Lett* 2007;**98**(20):206102.
41. Wigner E. On the interaction of electrons in metals. *Phys Rev* 1934;**46**:1002–11.
42. Keller DJ, McConnell HM, Moy VT. Theory of superstructures in lipid monolayer phase transitions. *J Phys Chem* 1986;**90**:2311–5.
43. Lukas S, Witte G, Ch W. Novel mechanism for molecular self-assembly on metal substrates: unidirectional rows of pentacene on Cu(110) produced by a substrate-mediated repulsion. *Phys Rev Lett* 2001;**88**:028301.
44. Nanayakkara SU, Sykes ECH, Fernndez-Torres LC, Blake MM, Weiss PS. Long-range electronic interactions at a high temperature: bromine adatom islands on Cu(111). *Phys Rev Lett* 2007;**98**:206108.
45. Silly F, Pivetta M, Ternes M, Patthey F, Pelz JP, Schneider WD. Creation of an atomic superlattice by immersing metallic adatoms in a two-dimensional electron sea. *Phys Rev Lett* 2004;**92**:16101.
46. Ternes M, Weber C, Pivetta M, Patthey F, Pelz JP, Giamarchi T, et al. Scanning-tunneling spectroscopy of surface-state electrons scattered by a slightly disordered two-dimensional dilute 'Solid': Ce on Ag (111). *Phys Rev Lett* 2004;**93**:146805.
47. Tersoff J, Hamann DR. Theory and application for the scanning tunneling microscope. *Phys Rev Lett* 1983;**50**:1998–2001.
48. Wennmohs F, Staemmler V, Schindler M. Theoretical investigation of weak hydrogen bonds to sulfur. *J Chem Phys* 2003;**119**:3208.
49. Glotzer SC, Stauffer D, Jan N. Monte Carlo simulations of phase separation in chemically reactive binary mixtures. *Phys Rev Lett* 1994;**72**:4109–12.
50. Sagui C, Desai RC. Kinetics of topological defects in systems with competing interactions. *Phys Rev Lett* 1993;**71**:3995–8.
51. Tomba G, Stengel M, Schneider W-D, Baldereschi A, de Vita A. Supramolecular self-assembly driven by electrostatic repulsion: the 1D aggregation of rubrene pentagons on Au(111). *ACS Nano* 2010;**4**:7545–51.

52. Anderson PW. Ground state of a magnetic impurity in a metal. *Phys Rev* 1967;**164**:352–9.
53. Nagaoka K, Jamneala T, Grobis M, Crommie MF. Temperature dependence of a single Kondo impurity. *Phys Rev Lett* 2002;**88**:077205.
54. Fano U. Effects of configuration interaction on intensities and phase shifts. *Phys Rev* 1961;**124**:1866–78.
55. Madhavan V, Chen W, Jamneala T, Crommie MF, Wingreen NS. Tunneling into a single magnetic atom: spectroscopic evidence of the Kondo resonance. *Science* 1998;**280**:567.
56. Gordon DG, Gres J, Kastner MA, Shtrikman H, Mahalu D, Meirav U. From the Kondo regime to the mixed-valence regime in a single-electron transistor. *Phys Rev Lett* 1998;**81**:5225–8.
57. Ho W. Single-molecule chemistry. *J Chem Phys* 2002;**117**:11033.
58. Kushmerick JG, Lazorcik J, Patterson CH, Shashidhar R, Seferos DS, Bazan GC. Vibronic contributions to charge transport across molecular junctions. *Nano Lett* 2004;**4**(4):639–42.
59. Park H, Park J, Lim AKL, Anderson EH, Alivisatos AP, Mceuen PL. Nanomechanical oscillations in a single-C60 transistor. *Nature* 2000;**407**:57.
60. Pascual JI, Herrero JG, Portal DS, Rust HP. Vibrational spectroscopy on single C60 molecules: the role of molecular orientation. *J Chem Phys* 2002;**117**:9531–4.
61. Stipe BC, Rezaei MA, Ho W. Single-molecule vibrational spectroscopy and microscopy. *Science* 1998;**280**:1732.
62. Wang W, Lee T, Kretzschmar I, Reed MA. Inelastic electron tunneling spectroscopy of an alkanedithiol self-assembled monolayer. *Nano Lett* 2004;**4**:643.
63. Flensberg K. Tunneling broadening of vibrational sidebands in molecular transistors. *Phys Rev B* 2003;**68**:205323.
64. Paaske J, Flensberg K. Vibrational sidebands and the Kondo effect in molecular transistors. *Phys Rev Lett* 2005;**94**:176801.
65. Lipari NO, Duke CB, Bozio R, Girlando A, Pecile C, Padva A. Electron—molecular-vibration coupling in 7,7,8,8-tetracyano-p-Quinodimethane (TCNQ). *Chem Phys Lett* 1976;**44**:236–40.
66. Rice MJ, Lipari NO, Strssler S. Dimerized organic linear-chain conductors and the unambiguous experimental determination of electron-molecular-vibration coupling constants. *Phys Rev Lett* 1977;**39**:1359–62.

Part II

Chemistry at the Atomic-Scale

Chapter 4

Imprinting Atomic and Molecular Patterns

Iain R. McNab and John C. Polanyi
Department of Chemistry and Institute of Optical Sciences, University of Toronto, Toronto, Ontario, M5S 3H6, Canada

1. SINGLE-MOLECULE IMPRINTING ON SEMICONDUCTORS

Single-molecule imprinting refers to the pattern of reactive outcomes stemming from the reaction of individual molecules at surfaces. In the following, we describe recent progress in using self-assembled molecules as templates for permanently chemically attaching patterns to surfaces. For a historical survey of patterned atomic reaction at surfaces, as studied by scanning tunnelling microscopy (STM), we refer the reader to our earlier review.[1]

STM has enabled a transformative development in the history of surface science, as it relates to surface reaction. There exists today the possibility of translating the rich field of molecular reaction dynamics from gases to surfaces. The new approach will, moreover, constitute a more powerful method than any for gases, as a single molecular reagent can, in principle and sometimes in practice, be observed in relation to the atoms of the substrate beneath, both before and after reaction occurs. This extreme level of detail is required for full understanding, as the surface is a vital participant in the reactive event. Used in this fashion, the STM can be expected ultimately to provide the essential instrument for solving the long-standing riddle of heterogeneous catalysis.

The power of STM as a tool for the study of atomic reaction at surfaces was apparent, to name a prominent example, in the early work of Avouris and Wolkow[2] who observed a preference for the reaction of NH_3 with adatom dangling bonds at the surface of Si(111)-7 × 7 in the sequence of reactivity; 'rest atom' greater than 'centre atom' which in turn was more reactive than the 'corner atom'. A brief description of the Si(111)-7 × 7 surface, employed in this pioneering study and other studies described below, is given in Figure 4.1. Some years would pass before such observations could be translated into the characterization of the molecular dynamics of surface reaction. For ammonia, the interpretation is still incomplete, but it is known that at

Frontiers of Nanoscience, Vol. 2. DOI: 10.1016/B978-0-08-096355-6.00004-0

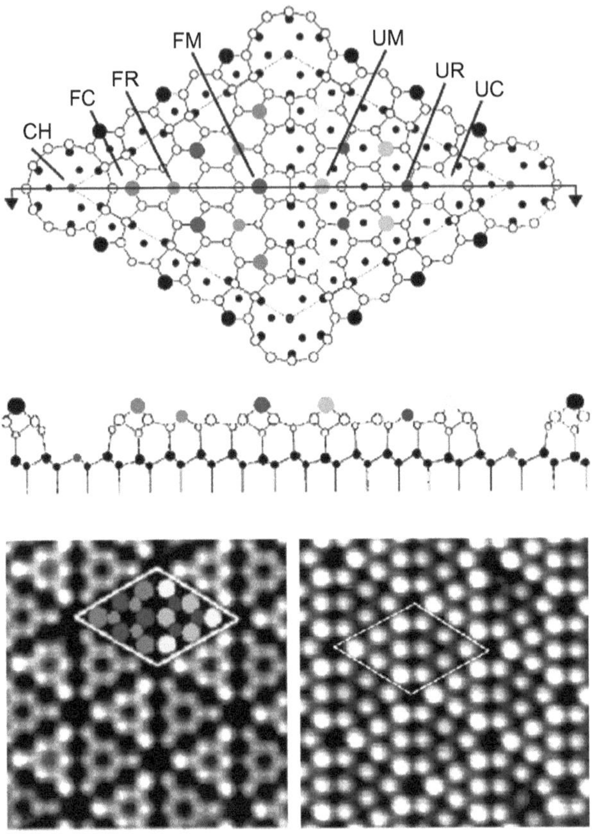

FIGURE 4.1 The Si(111)-7 × 7 surface reconstruction. (Top) The dimer adatom stacking fault model of the reconstruction[3] contains seven types of atoms that are inequivalent and have dangling bonds. Each of these types is present on the centre section of the unit cell. They are, from left to right in the schematic section, Corner Hole (CH) adatom (dark green), Faulted-half Corner (FC) adatom (red), Faulted-half Rest (FR) atom (light purple), Faulted-half Middle (FM) adatom (blue), Unfaulted-half Middle (UM) adatom (light green), Unfaulted-half Rest (UR) atom (dark purple) and Unfaulted-half Corner (UC) adatom (yellow). The same colour coding has been superimposed on one unit cell of the STM image beneath. (Adapted from Brommer *et al.*[4], copyright (1994), with permission from Elsevier.) (Bottom left) An STM image of the Si(111)-7 × 7 reconstruction taken in negative bias(V_{surf}=-1.5 V, I_{tun}=0.41 nA). This topographic image is unusual in showing both rest atoms and adatoms, The faulted- (bright) and unfaulted- (less bright) halves of the unit cell are clearly distinguishable. Also clearly distinguished is the difference between faulted centre (FC) and faulted middle (FM) adatoms, and also between unfaulted centre (UC) and unfaulted middle (UM) adatoms. Midway between the corner adatoms and the middle adatoms are the rest atoms (FR and UR). (Bottom right) A more usual STM image of the Si(111)-7 × 7 surface taken in negative bias (V_{surf} = −0.57 V, I_{tun} = 0.3 nA), clearly showing the adatoms in the two halves of the unit cell; the faulted half is bright, the unfaulted half dark. A unit cell with sides of length 26.9 Å is outlined in each image. (STM Topographs reprinted with permission from Wang *et al.*[5] Copyright (2004) by the American Physical Society. (For interpretation of the references to colour in this figure legend, the reader is referred to the Web version of this chapter.)

room temperature NH_3 dissociatively adsorbs to give NH_2+H with the NH_2 primarily bound to the rest-atom sites.[6]

Knowing the positions of the reagent and product species in the (virtually) two-dimensional world of the surface, the intermediate motions and the forces that govern them can be inferred with some confidence. This information constitutes the molecular dynamics of surface reaction, studied one molecule at a time by STM.

For the case of ammonia reacting at Si(111), this still constitutes a challenge at the time of writing though the outlines of the solution to the problem of the 'molecular dance' in this reaction have begun to emerge from experiments in several laboratories,[7,8] supported by a number of *ab initio* computations based on density functional theory. The findings will be summarized in Section 1.3.

1.1. Localized Atomic Reaction

Localized atomic reaction (LAR) is the term introduced[9] to describe the reaction of a molecule at a surface giving rise to a chemisorbed imprint of product closely adjacent to the original position of the reagent. LAR has the valuable attribute that it enables patterns of physisorbed molecules, created by self-assembly, to be rendered permanent by chemical reaction *without* destroying the pattern; it suggests, therefore, a method of 'molecular printing'.

The identification of LAR was made for the electron-induced reaction of chlorobenzene at the Si(111)-7 $\times$ 7 surface, induced by the tunnelling current from an STM tip.[9,10] Although it was not known at that date, chlorobenzene attaches to the Si(111)-7 $\times$ 7 surface by creating two sigma bonds with a pair of silicon atoms (silicon adatom and silicon-rest atom, adjacent to one another) at the surface.[11–13] Electron-induced reaction gave rise to a localized Cl—Si product, as is evident in Figure 4.2,[9] in which sequential electron pulses (+4 V at the surface) were applied to a Si(111)-7 $\times$ 7 covered with a monolayer of chlorobenzene. In the image at the left, the pulsing was continuous as the tip moved 250 Å down the *y*-axis, whereas in the right-hand image, the tip was pulsed at 60 Å intervals along the *y*-axis. The bright features give the location of the Cl atom reaction product bound to the silicon surface. The reaction product is seen to be localized ($\pm$ order of 10 Å) to the site of the electron-induced reaction. This constitutes molecular 'writing'. The same study exemplified 'molecular printing' of a previously self-assembled pattern by applying current from the tip in a random fashion to a surface on which $59\pm1\%$ of the molecules had assembled on the faulted half of each unit cell, and $41\pm1\%$ on the unfaulted half; the Cl reaction product mirrored this reagent distribution. The observed LAR was ascribed to stabilization of the reactive transition state (TS) by the localized coexistence of the C—Cl bond being broken and the Cl—Si bond being formed. *Ab initio* calculations of atomic reaction of chlorine from chlorobenzene on the Si(111)-7 $\times$ 7

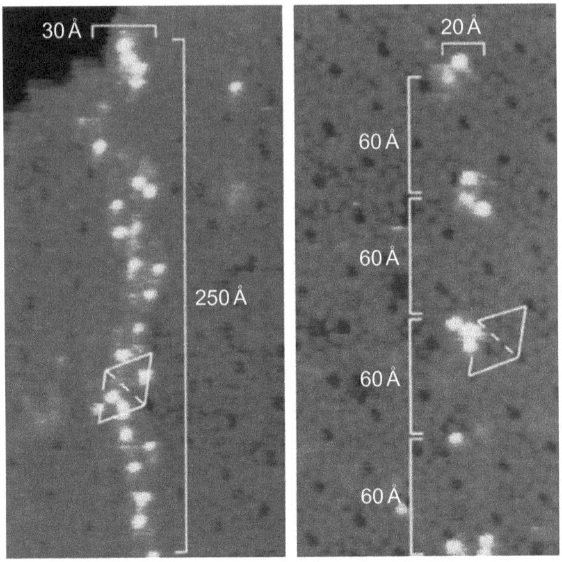

FIGURE 4.2 STM images ($V_{surf}=-3$ V, $I_{tun}=0.09$ nA) of the Si(111)-7 × 7 surface with chlorine atoms (bright) imprinted from chemisorbed chlorobenzene. Chlorine atoms were imprinted in (A) as a continuous line by application of a rapid sequence of pulses to the STM tip which was moved in the y-direction and in (B) as clusters of one to three chemisorbed chlorine atoms after application of voltage pulses ($V_{surf}=+4$ V) while the tip was moved by 60 Å to successive positions along a line in the y-direction, extending from top to bottom in the figure. A unit cell with sides of length 26.9 Å is outlined in each image. Adapted with permission from Ref. 9. (A colour version of this figure is available in the Web version of this chapter.)

surface gave further evidence that the transfer of the chlorine atom from the chemisorbed chlorobenzene molecule to the dangling bond of a nearby silicon adatom occurred by concerted reaction.[9,14]

In a detailed experimental study of the electron-induced reaction of chlorobenzene with Si(111)-7 × 7, Sloan and Palmer[15] showed that this was a two-electron process. More recent work from the same laboratory[16] raised the possibility that the first electron may have converted chemisorbed chlorobenzene into physisorbed chlorobenzene (a conversion achieved thermally in Ref. 16), whereupon the second electron entering an antibonding orbital was sufficient to cause reaction. Localized reaction predominated, particularly at low STM currents, but a minority of delocalized events were observed.[10] The FWHW of distribution of the observed Cl atom distribution at room temperature[10] was ~10 Å at a tunnelling current, I_{tun}, of −0.1 nA, ~20 Å at $I_{tun}=-0.5$ nA.

The LAR of CH_3Br on Si(111)-7 × 7 has been studied by experiment and theory.[17] For this process, the initial state (IS) was physisorbed methyl bromide at 50 K, which self-assembled into circles due to a preference for methyl physisorption at the 'middle' adatom sites on the Si(111)-7 × 7 surface. Reaction to brominate the adjacent corner adatom was induced by either 193 nm

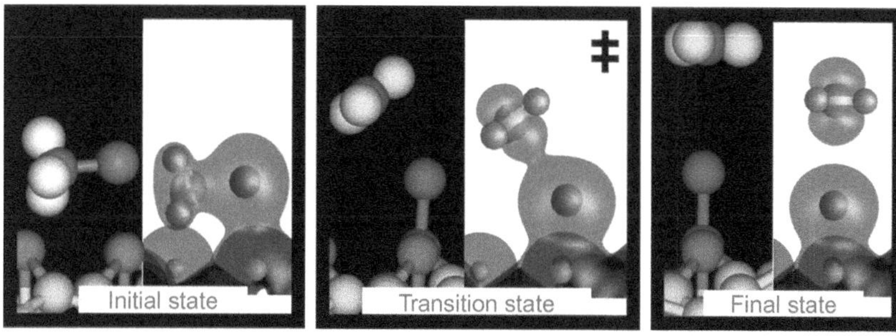

FIGURE 4.3 DFT calculation of the stages of localized surface bromination of Si(111)-7 × 7 by methyl bromide (methyl at left, Br at right in IS). Side views of the thermal reaction path showing the initial state (IS), transition state (TS) and final state (FS). At left of each panel is a ball and stick model while the right of each panel shows atom positions and isosurface contours of charge densities set to 0.05 e/Å^3. The computed activation barrier for thermal reaction, with its crest at position 7 (the transition state ‡), was 0.32 eV. Reproduced with permission from Ref. 17. (A colour version of this figure is available in the Web version of this chapter.)

irradiation or electrons of energy $E_0 > 1.86$ V.[17,18] In either case, localized reaction formed circles of chemisorbed Br atoms, where previously circles of physisorbed CH_3Br(ad) had been observed. The CH_3 radical was presumed to be released into the gas phase, as it was never observed at the surface. *Ab initio* calculations gave evidence of concerted reaction in successive stages from IS to final state, as shown in Figure 4.3.

It should be stressed that the pattern of reaction shown in Figure 4.3 in which a horizontal CH_3Br tilts its methyl away from the surface in the TS (central panel; double-dagger designation) is characteristic of this adsorbate's reaction dynamics on a particular surface, Si(111)-7 × 7. In Section 1.3, we shall describe the reaction dynamics of the same molecule on a different face of silicon, namely, Si(100)-2 × 1. On this latter surface, Si(100), methyl bromide invariably reacts by dissociative attachment, chemisorbing so as to attach both CH_3 and Br at adjacent Si sites.

1.1.1. Molecular-Scale Imprinting

In the preceding section, we reviewed the building blocks necessary for molecular-scale imprinting (MSI). A pattern is created at a surface by self-assembly of mobile molecules in a physisorbed state. LAR is then induced to chemisorb a component of the adsorbed molecules (a halogen atom in the examples cited) in a pattern that closely resembles the initial physisorbed pattern. The latter is the 'imprinting' stage in which the mobile physisorbed species is immobilized by chemisorptions. The fidelity with which the physisorbed pattern is matched by the chemisorbed one is a measure of the degree of 'localization' of the chemical reaction. The chemical reaction ('imprinting') can be induced by means of heat, electrons (or other charged particles) or light. Harikumar *et al.*[19] compared all three instrumentalities for a single system. The reason that the induced reaction is often localized is thought to be that the lowest energy pathway to reaction is a 'concerted' reaction in which bond breaking and bond formation occur concurrently, as pictured by *ab initio* calculation for physisorbed CH_3Br brominating Si(111)-7 × 7 in Figure 4.3. Simultaneous bond rupture and bond formation require localization.

A simple model calculation was used previously to test the hypothesis that concerted reaction would minimize the energy barrier for halogenations at a silicon surface.[20] In this work, the energy barrier to dihalogenation of a model silicon surface by *p*-dibromobenzene was calculated for numerous possible Br atom separations at the surface. The lowest energy barrier was obtained for localized reaction, with the reacted (chemisorbed) bromine atoms at the surface only 3–4 Å further apart than they had been in the physisorbed reagent molecule. In concerted reaction, coexistence of the 'old' and the 'new' chemical bonds (that being broken and that being formed) can be expected to add to the stability of the TS, lowering the activation energy. Coexistence in time requires proximity in space. This suggests that at reagent energies not greatly in excess of that required to induce reaction, localized reaction will be the norm.

These considerations have led to a variety of proposals for MSI, all involving self-assembly of a physisorbed pattern followed by imprinting by LAR. The processes were detailed in a US Patent (2000)[21] and elaborated in later patents.[22–24]

1.2. Single Dissociative Attachment 1: Diatomic Molecules at Si(100)-2 × 1

We define single dissociative attachment as a dissociative attachment even in which a single bond is broken in a physisorbed species. We first consider single dissociative attachment for diatomic molecules and then for polyatomic molecules.

The dissociative attachment of diatomic molecules at Si(100)-2 × 1 can occur in three principal patterns, which we refer to as OD (on-dimer), ID (inter-dimer) and IR (inter-row), see Figure 4.4 (bottom) which gives a brief introduction to the room temperature Si(100) surface. There is a notable lack of simple empirical or theoretical guides to the reaction path in dissociative attachment. We begin by describing three cases of dissociative attachment of simple diatomic molecules: H_2, Cl_2 and O_2. No physisorbed precursors were observed prior to chemisorption. Nevertheless, such states are likely to exist but be short-lived.

1.2.1. Dissociative Attachment of H_2 at Si(100)-2 × 1

The dissociative attachment of H_2 at Si(100)-2 × 1 was recently reviewed.[26,27] Dissociative attachment occurs ID at two adjacent silicon-dimers of a single silicon dimer row, with a final H···H separation of 3.8 Å, in preference to dissociative attachment OD, giving a final H···H separation of 3.5 Å. There is clearly a large mismatch between the bond length of molecular hydrogen at 0.74 Å and the final H···H separation of 3.8 Å. The ID dissociative

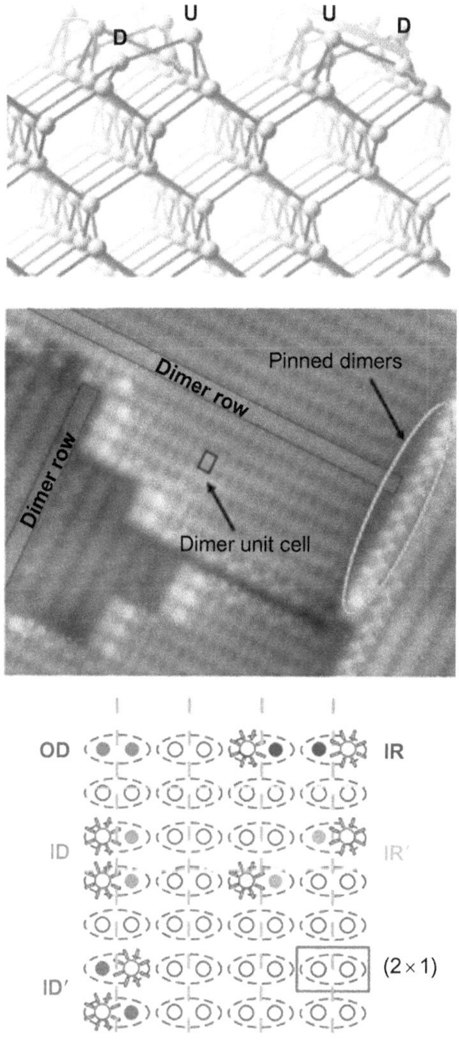

FIGURE 4.4 The Si(100) silicon surface. (Top) Ball and stick model showing 'buckled' silicon dimers (yellow) in their staggered positions corresponding to a frozen, or pinned, surface with c (4 × 4) reconstruction. The silicon dimers alternate in their tilt orientation (D, 'down' silicon atom; U, 'up' silicon atom; a pink rectangle marks the centre of one silicon dimer row). (Centre) A room temperature STM image of a Si(100) surface without adsorbates or defects ($V_{surf} = -2$ V, $I_{tun} = 0.2$ nA, 180 Å × 130 Å). On the terraces, at temperatures above 120 K, each silicon dimer flips between up-down and down-up, at a rate of order 10^{11} s^{-1}, which is rapid on the timescale of STM, and so silicon dimers are imaged in their *average* positions as a Si(100)-2 × 1 reconstruction. A single silicon dimer 2 × 1 unit cell (7.68 Å × 3.84 Å) is outlined in black, and the direction of two representative silicon dimer rows on adjacent terraces is indicated. Close to step-edges, silicon dimers are 'pinned' and can be seen in the buckled configuration (one such region is outlined in green) that corresponds to the ground state Si(100)-c(4 × 2) reconstruction into which the

attachment is less exothermic than OD but is nevertheless preferred, having the lower energy barrier. Subsequent H atom migration allows the lower-energy OD state to be achieved over time, even at a room temperature Si (100) surface.[28] From the interatomic spacing in H_2 ($r_o=0.74$ Å), one might expect to find the lowest energy barrier to DA for a OD addition (Si···Si separation of 2.7 Å) rather than ID (ID=4.0 Å), yet ID attachment has been shown both experimentally[29] and theoretically[30–32] to have the lower barrier. We revert to this example later (Section 1.4).

1.2.2. Dissociative Attachment of Cl_2 at Si(100)-2 × 1

As a second simple system, we consider the dissociative attachment of Cl_2 at Si(100)-2 × 1.[25] The final states were pairs of chemically attached Cl—Si atoms found distributed as 15% OD, 33% ID and 52% IR. Once again, IR final states predominate, despite the large mismatch of distance between the internuclear separation of 1.99 Å and the available Si···Si separations of 2.7 Å (OD), 4.0 Å (ID) and 5.2 Å (IR).

1.2.3. Dissociative Attachment of O_2 at Si(100)-2 × 1

Dissociative adsorption of O_2 on Si(100)-2 × 1 has been studied for over 40 years, but there have been few studies at low coverage so there is not yet a definitive atomic level mechanism. Recently, a specific metastable 'silanone' intermediate species has been identified at a defect-free surface of Si(100)-2 × 1 by infrared spectroscopy.[33] This has been confirmed by both low temperature STM and *ab initio* calculation.[34] The 'silanone' species is formed by dissociative attachment at a single silicon dimer. It consists of one oxygen atom inserted into the silicon-dimer bond, while the remaining oxygen atom attaches to one of the two available dangling bonds of the same silicon dimer, a variant, therefore of the OD reaction pathway.

1.3. Single Dissociative Attachment 2: Dissociative Attachment of Polyatomics

The few examples of dissociative attachment of polyatomics so far studied in detail have proved difficult to analyze, often requiring significant inputs from

---

surface freezes at temperatures below 120 K, and which is modelled at top (previously unpublished work). (Bottom) A schematic representation of the Si(100)-2 × 1 surface, showing the most commonly found patterns of dissociative attachment: on dimer (OD), inter row (IR), inter row diagonal (IR'), inter dimer (ID), inter dimer diagonal (ID′). Silicon dimers are represented by dashed ovals, individual silicon atoms by solid circles and adsorbate pairs by filled coloured circles. The centre of silicon dimer rows are shown by pink dotted lines and the yellow sunbursts represent dangling bonds, formed during chemisorption at a single silicon atom of a silicon dimer, which are imaged brighter than the clean substrate in STM. A single 2 × 1 unit cell is outlined with a blue rectangle. Courtesy of S. Y. Guo, following Ref. 25. (For interpretation of the references to colour in this figure legend, the reader is referred to the Web version of this chapter.)

different groups of experimenters and theorists, before a consensus was achieved. This is exemplified by the case of the dissociative attachment of ammonia at Si(100)-2 × 1 briefly discussed below. As a further example, we then consider the DA of water at this same surface.

1.3.1. Ammonia at Si(100)-2 × 1

The literature on the dissociative adsorption of NH_3 at Si(100)-2 × 1 is long and controversial. It was recently reviewed by Owen (who contrasted the behaviour with PH_3 at Si(100)-2 × 1)[6] and, from a different perspective, by Perrine and Teplyakov.[35] The current understanding is that at room temperature and below, NH_3 adsorbs in a short-lived precursor state arising from a dative bond between the lone electron pair on the N atom and a 'down' silicon atom of a silicon dimer. At room temperature, reaction is rapid so that only reacted products were observed, but at 65 K, both precursor and products were observed simultaneously[36] on the Si(100)c(4 × 2) reconstructed surface (the surface that results once silicon-dimer flipping is frozen out). Reaction products were shown by STM to include both OD and ID adsorption of H and NH_2.[37] The best available *ab initio* calculations[38] give energy barriers to dissociative attachment of 0.87 eV (OD), 0.77 eV (ID) and 1.12 eV (IR). Interestingly, this is the converse of the order found for methyl bromide dissociative attachment at the same surface (see Section 1.3.3). Experimentally, IR adsorption has not been observed. For low surface coverage, (0.01 L) the ratio of OD to ID was found to be 0.75%:1.00%, in qualitative agreement with the calculated energy barriers. However, as coverage was increased, this ratio reversed, becoming 3.5%:2.5% by 0.04 L,[37] suggesting, interestingly, that even at these low coverages, prior adsorption at the surface significantly effects subsequent dissociative attachment. Adsorption of ammonia onto Si (100) 2 × 1, in both physisorption (NH_3) and chemisorption ($NH_2 \cdots H$), causes charge redistribution within the surface.[39,40] The redistribution of charge would seem to steer the course of further physisorption/chemisorption events, resulting in a wealth of co-adsorption structures and patterning[41]; the interested reader is referred to Owen for a complete discussion.[6]

Although the reaction pathways for dissociative attachment of NH_3 at Si (100)-2 × 1 are now understood, the assignment of individual features in the STM images remains in contention; there is not yet a consensus as to which features in the STM images are due to chemisorbed NH_2—Si and which due to chemisorbed H—Si (see Ref. 6 and references therein).

1.3.2. Water at Si(100)-2 × 1

Water is one of the enemies of ultra-high vacuum (UHV) and is not commonly introduced into UHV systems deliberately. The current understanding of the dissociative attachment of H_2O at Si(100)-2 × 1 is a history of the identification, and subsequent manipulations, of the 'C-type defect' of the

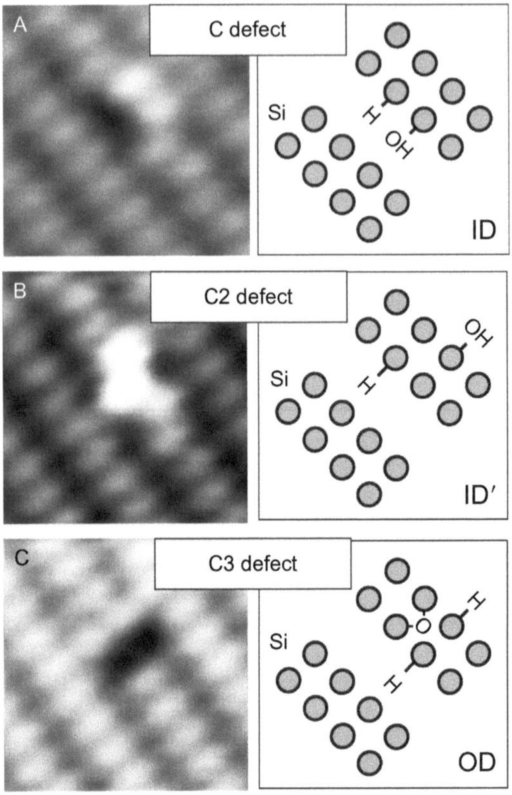

FIGURE 4.5 C defects on Si (100)-2×1. (A) C defect (inter-dimer, ID) can be transformed to (B) the C2 defect (inter-dimer diagonal ID') and (C) the C3 defect (on dimer; OD). The images were chosen from long-time observations of a *single* defect. Reversible switching C↔C2 finally resulted in formation of the stable C3 defect consisting of two H-atoms attached On Dimer and the O-atom in a bridging position on the adjacent dimer. Schematic drawings on the right side show positions of the bonded H atom, O atom and OH group. Filled state images, $V_{surf} = -2$ V, $I_{tun} = 0.6$ nA, 17 Å × 17 Å. Adapted with permission from Ref. 47. (A colour version of this figure is available in the Web version of this chapter.)

Si(100)-2×1 surface. Early STM images of Si(100)-2×1[42] revealed three common defects. 'A-type' and 'B-type' defects were quickly identified as being missing silicon atoms, but the third 'C-type' defect was more difficult to identify. Independently, two groups showed that 'C-type' defects were water molecules that had dissociatively attached at adjacent silicon dimers in an ID fashion to form H—Si and OH—Si, at adjacent dimers together with two dangling bonds.[43,44] Further observations showed an additional defect, C2[45] which can undergo reversible transitions with C defects, also forming C2–C2 structures.[46] These structures, their reversible transformations with C-defects together with still further structures, were subsequently assigned and interpreted as due to movements of the OH group under the influence of the STM tip[47] ,the final stable configuration C3 consists of two H-atoms attached On Dimer, and the O-atom attached in a bridging position on an adjacent dimer (see Figure 4.5).

1.3.3. Methyl Bromide at Si(100)-2×1

Recent work reports the dissociative attachment of CH_3Br at room temperature and below, employing STM interpreted by *ab initio* theory.[48] It was found that CH_3Br dissociatively attached to form principally IR outcomes (88.5%) and subsidiarily ID (10.5%) and OD (1%). *Ab initio* theory predicted three most stable physisorbed precursor states with similar heats of adsorption leading individually to these three chemisorbed outcomes with calculated classical barrier heights, Ec, of 0.48 eV (IR), 0.58 eV (ID) and 0.63 eV (OD). The physisorbed states differed in the extent to which they positioned the CH_3Br reagent preferentially on top of a dimer pair (leading largely to OD reaction), or with the C—Br axis along a dimer row (leading mainly to ID reaction) or with the molecule lying midway between rows (leading to IR reaction, see Figures 4.6 and 4.7). The observed multiple outcomes of dissociative attachment may therefore occur through differing physisorption geometries. These, in turn, will be determined by different ways of maximizing the interaction of a molecule, of specific length and dipole moment, with the complex features of the underlying surface. For CH_3Br, reactions from the three major physisorbed states give in one case IR, in another OD and

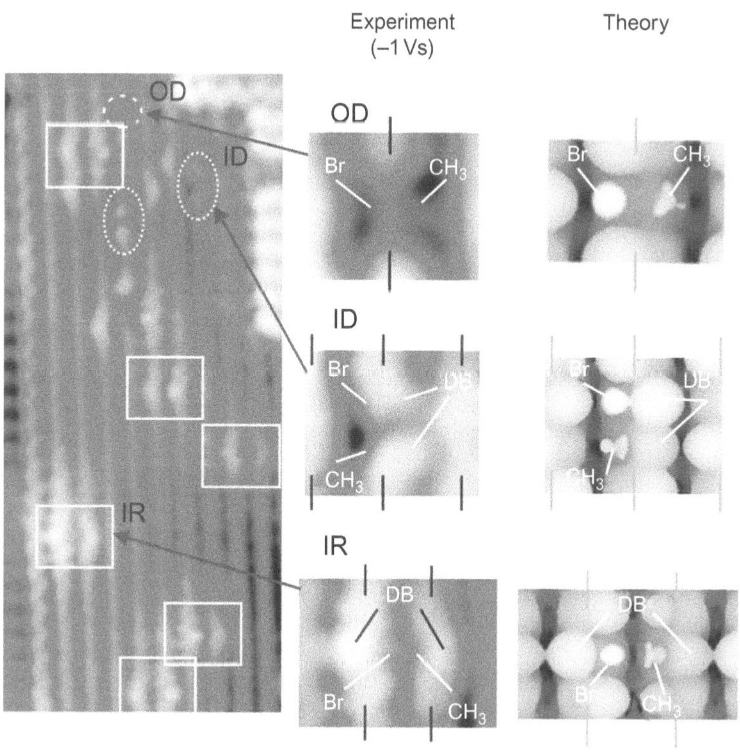

FIGURE 4.6 Reaction outcomes for methyl bromide on Si(100)-2 × 1 imaged by STM topographs (V_{surf} = -1 V, I_{tun} = 0.2 nA, 90 × 200 Å^2). In the wide area image (left column), three types of feature are observed, 'on-dimer' (OD, 1%), 'inter-dimer' (ID, 10%) and 'inter-row' (IR, 89%). Close-up high-resolution images of each feature are shown (middle column) together with theoretically simulated STM topographs of each feature (right column). (A colour version of this figure is available in the Web version of this chapter.)

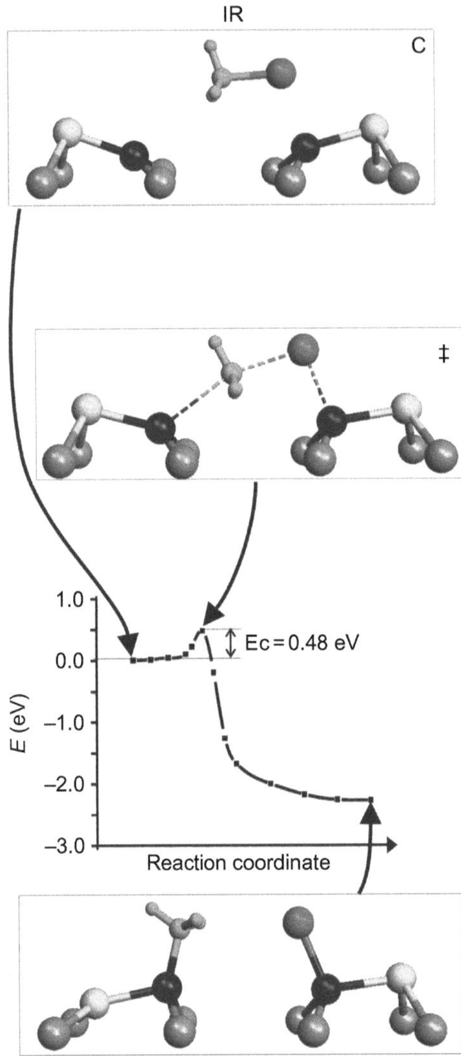

FIGURE 4.7 *Ab initio* calculations of the principal, inter-row (IR) reaction of methyl bromide at Si(100)-2 × 1, showing the initial physisorbed state, transition state and final state, and a 'nudged elastic band' calculation of the minimum energy path for reaction. (A colour version of this figure is available in the Web version of this chapter.)

in the third ID. Differences between the computed activation energies for dissociative attachment agreed satisfactorily with the measured differences in activation energy for these three outcomes. The overall outcome in this example appeared, therefore, to be determined principally by the differences in activation energy for the three dissociative pathways.

1.3.4. Methyl Chloride at Si(100)-2 × 1

The dissociative attachment of the CH_3Cl system is to be found reported in a later section (Section 4.3), as it is an example of ordered chain reaction at silicon. For CH_3Br, chain formation represents only a minor reaction pathway.

Two factors may mitigate against chain formation for CH_3Br, despite the presence at the surface of a large number of dangling bonds formed in the major reactive pathway of IR and in the lesser pathway of ID. The first factor is the high reactivity of CH_3Br with the surface as compared with CH_3Cl (C—Br being a weaker bond than C—Cl), with the result that incoming CH_3Br molecules react individually by dissociative attachment rather than roaming the surface as required for chain formation. A second factor could be lesser mobility for physisorbed CH_3Br (with its large polarizability) than for CH_3Cl.

1.4. Discussion of Single Dissociative Attachment

We have seen that physisorption geometries constitute the ISs for the reactive pathways to dissociative attachment. As they are dependent upon an identifiable catalogue of reagent properties (size, shape, polarity and polarizability of the adsorbate and its underlying region of substrate), they can be computed. This constitutes, however, only the first step on the path to dissociative attachment which involves crucially the geography of the reactive potential-energy 'surfaces' leading from the various physisorbed ISs across TSs to the chemisorbed reaction product. The probabilities of the various dissociative attachment outcomes may be influenced by the relative yields of physisorbed states but in addition depend sensitively on the height of the energy barriers to reaction from these physisorbed states. General predictive guides to dissociative attachment pathways are today lacking. As, however, both experiment and theory are now within reach, this constitutes a contemporary challenge. In the meantime, we can expect that as the potential-energy surface (PES) is a continuous function, for related dissociative attachment reactions, greater exothermicity for a reaction path will correlate with lower energy barriers.

Recently, it was suggested (see Figure 4.8) that the *location of the energy barrier* to dissociative attachment along the reaction path correlates with the preferred separation of the product fragments when chemisorbed at the surface; 'early' barriers leading to attachment of the fragments close to one another, and 'late' barriers to attachment at more widely separated locations on the surface.[49] This, in turn, (see Figure 4.9) suggested a means for steering dissociative attachment reaction using reagent translation to yield products closer together, or reagent vibration to place them further apart.

1.5. Multiple Dissociative Attachment

In the examples in Sections 1.1 and 1.2, one bond broke in the adsorbate while two bonds formed on the surface. We use the designation 'multiple' dissociative attachment for cases in which two (double dissociative attachment) or more bonds break in the adsorbate. We exemplify this by an early study of the severing of two C—Br bonds in dibromobenzene to form two Si—Br bonds at Si(111)-7 × 7.[50]

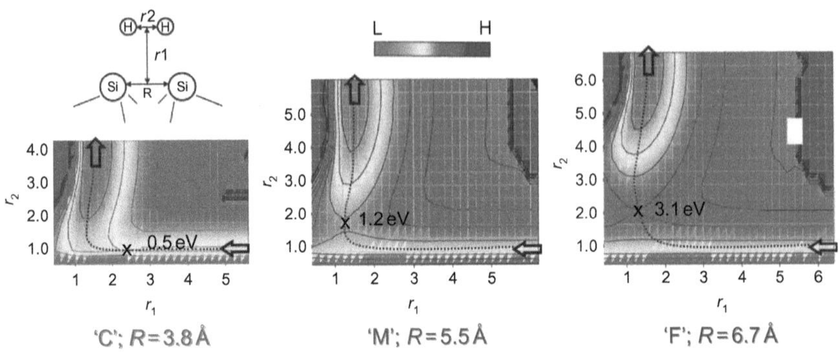

FIGURE 4.8 Three PESs calculated using the disilyl model, one for the 'close' (C) separation of the product H—H atoms, one for the 'medium' (M) separation and one for the 'far' (F) separation. The barrier can be seen to increase progressively and move to a later location in going from a H—H separation that is close to medium and then to far. The three PESs exhibit 'early', 'intermediate' and 'late' barrier crests with increasing adduct separation. Adapted with permission from Ref. 49. (A colour version of this figure is available in the Web version of this chapter.)

1.5.1. Double Dissociative Attachment of Dibromobenzenes

This double dissociative attachment was instructive, as it revealed a simple correlation between the reagent and product geometries (see Figure 4.10). On average, the 1,2-dibromobenzene was shown by STM to 'imprint' (i.e. chemisorb) pairs of Br—Si that were significantly more closely spaced than the pairs of Br—Si imprinted at the same surface by 1,4-dibromobenzene (see Figure 4.11). It was evident that greater separation between the halogen atoms in the reagent molecule led to greater mean separation between the product halogen atoms, bound to the surface.

1.5.2. A Simple Model of Dibromobenzene at Si(111)-7×7

A simple theoretical model[20] shed light on the correlation between atomic separation between substituents, Br, in the 1,4-dibromobenzene reagent molecule and the separation between the atoms Br—Si following chemisorption reaction, that is, in the process we term 'imprinting'. In this ultra-simple model of the silicon substrate, the dangling bonds at the surface were represented by a pair of silyl radicals, placed a distance R apart (see Figure 4.12). The computed height of the barrier to halogenation of this model 'surface' by 1,4-dibromobenzene was then plotted as a function of R. The barrier was found to go through a pronounced minimum at a separation designated R^* (see Figure 4.13). The reaction showed a preference for imprinting its halogen atoms, X, at a separation which was dependent on the X···X separation in the reagent molecule, but which exceeded the separation in the reagent by (a modest) 3–4 Å. The outcome constitutes a further example of 'localized' reaction, suggesting the possibility of using a single adsorbate molecule as a template for imprinting atomic-scale

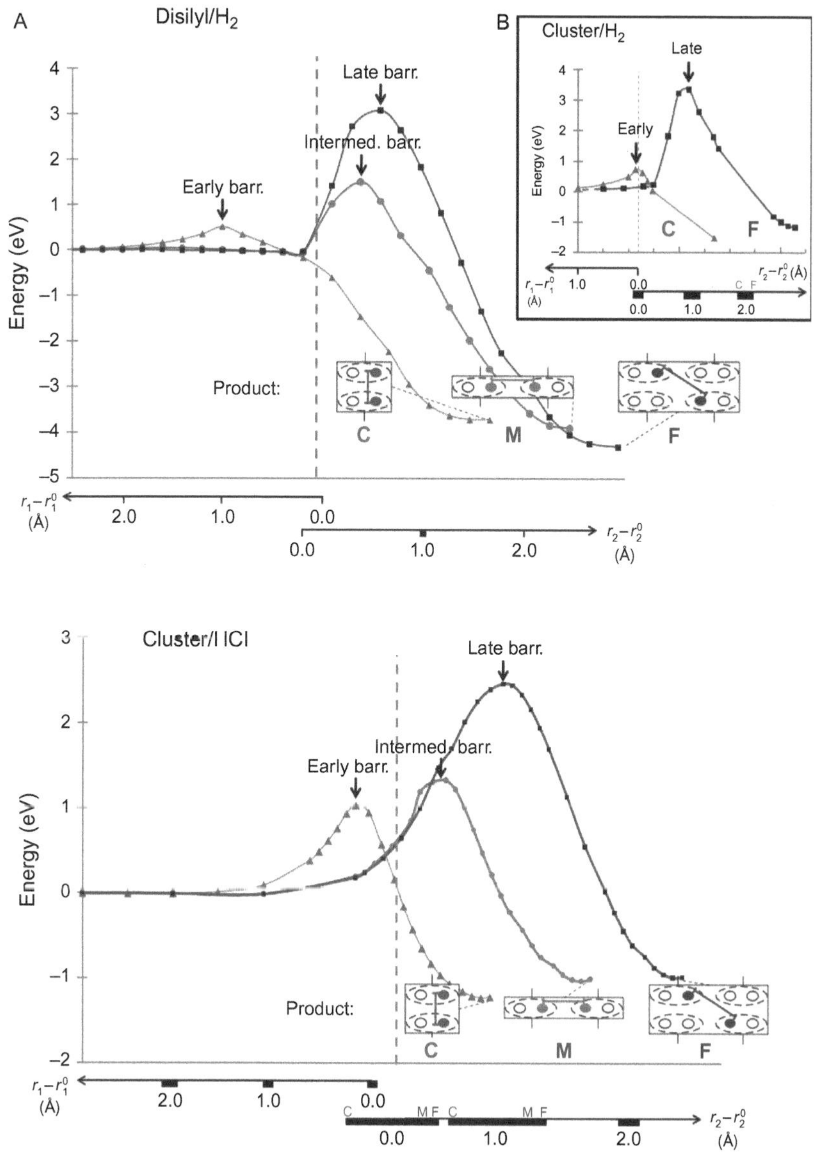

FIGURE 4.9 (Left) Energy profiles along the minimum energy paths for H_2 dissociative attachment at close (C), medium (M) and far (F) Si—Si separations (A) for the disilyl model and (B) for the $Si_{53}H_{44}$ cluster model. The dashed line indicates the point along the reaction coordinate at which the entry valley of the PES can be regarded as giving way to the exit valley; old and new bonds are equally extended, $r_1=r_1^0$, $r_2=r_2^0$. (Right) Energy profiles along the minimum energy paths for HCl dissociative attachment at close, medium and far Si—Si separations for the cluster model. Due to differences in the MEPs, positions along the reaction coordinate correspond to different values of the separation coordinate in the three cases C, M and F, as indicated by the horizontal black range of abscissa values. Adapted with permission from Ref. 49. (A colour version of this figure is available in the Web version of this chapter.)

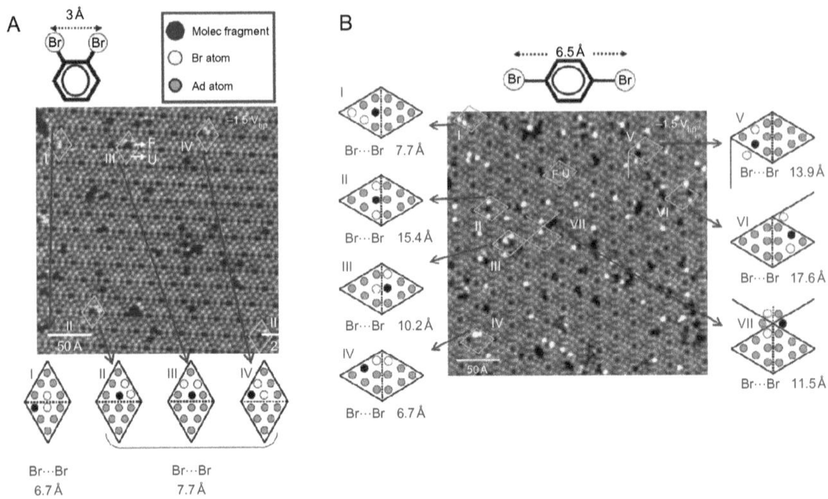

FIGURE 4.10 STM images ($V_s=+1.5$ V, $I_{tun}=0.2$ nA) of the Si(111)-7 × 7 following room temperature dissociative chemisorption of (A) 1,2-dibromobenzene and (B) 1,4-dibromobenzene. The bright yellow features are chemisorbed Br atoms, and the adjacent dark features are due to the organic residue. Green diamonds indicate unit cells (F, faulted; U, unfaulted). The reaction products (black and yellow) are indicated in the schematic diagrams (I–V for A; I–VII for B) with the same orientations as in the accompanying STM images. Histograms of the measured distributions of Br···Br pair distances are shown in the following figure. Numerous dark features not associated with bright spots are undissociated 1,2-diBrPh molecules. A unit cell with sides of length 26.9 Å is outlined in each image. Adapted with permission from Ref. 50. (For interpretation of the references to colour in this figure legend, the reader is referred to the Web version of this chapter.)

patterns of reaction product. In effect, the reacting molecule becomes a 'rubber stamp', of the type once favoured by officialdom. In the following section (Section 2.1), we extend this finding to the localized halogenation of a metal surface, imprinting pairs of iodine atoms at controlled separations on copper. A more apt analogy in this instance is with molecular callipers that, by selection of the reagent molecule, can be set to span different separations at the surface.

2. SINGLE-MOLECULE IMPRINTING ON METALS

As adsorbates migrate readily across the smooth surfaces of metals at room temperature and above, few studies exist of demonstrably *initial* distribution of reaction products from thermal reactions at metal surfaces.

2.1. LAR on Metals

Early attempts to ascertain the degree of localization of thermal reactions at metals began with an STM study of the dissociative attachment of O_2 at room

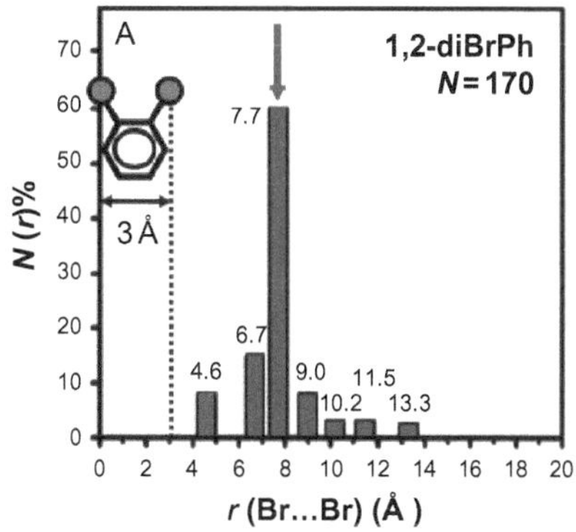

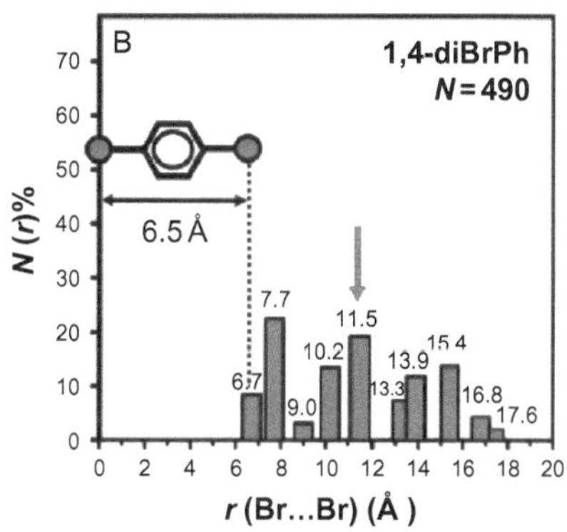

FIGURE 4.11 Distribution of Br atom pair separations (present as Br—Si) (A) for 1,2-dibromobenzene, (B) for 1,4-dibromobenzene. The magnitude of the pair separations is given, in Angstroms, above each histogram bar. The total number of Br atom pairs (N) is indicated for each histogram. The separation between the Br atoms in the intact molecule is indicated in each case. The weighted mean separation for each distribution is indicated with a downward arrow. Adapted, with permission, from Ref. 50. (A colour version of this figure is available in the Web version of this chapter.)

temperature Al(111).[51,52] Direct dissociation was regarded as responsible for the observation of pairs of O atoms over 80 Å apart. More recent work suggests that the widely separated O atoms originated in the abstraction of single O atoms in separate reactive events.[53] Dissociative chemisorption of O_2 on Al (111) gave rise to unresolved features in the STM images that were not larger than three lattice spacings, with a measured length (FWHM) of around 10 Å. The reader is referred to[1] for a more detailed discussion.

Thermal dissociation of O_2 at Pt(111) led[54] to the formation of O atom pairs separated on average by two lattice constants, that is, 5.55 Å[55] (see Figure 4.14). As the O—O separation in molecular oxygen is only 1.21 Å,[56] the observed separation in the chemisorbed state was ascribed to short-range 'transient ballistic motion'.

Electron-induced reaction, in contrast to thermal reaction, can be studied at reduced surface temperatures at which the complication of thermal diffusion is suppressed. Once again O_2 on a metal, Pt(111), was the first case studied.[57] Electron-induced reaction occurred at 50 K and was ascribed to vibrational ladder climbing at lower electron energies and direct resonant excitation of the O_2 at higher electron energy. In either case, the O atoms

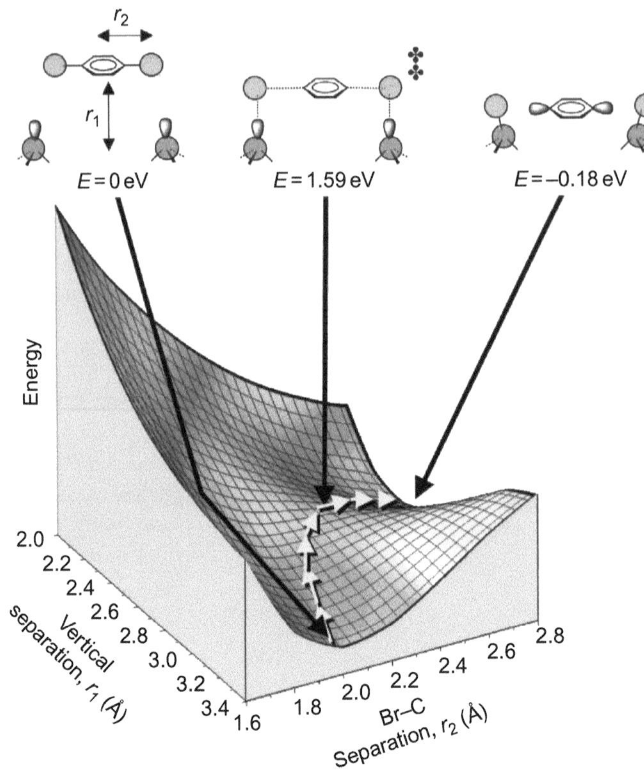

FIGURE 4.12 An example of a PES obtained in this study for a fixed Si···Si separation. The potential energy is plotted against r_1 and r_2 defined at top left. The initial, transition and final states are illustrated above and indicated on the PES. The PES is plotted for a *symmetric* stretch of the Br—C bonds. Reproduced, with permission, from Ref. 20. (A colour version of this figure is available in the Web version of this chapter.)

recoiled by one to three lattice constants (2.8–8.3 Å), much as previously reported for thermal dissociation.

The question of the degree of localization of surface reaction at metals has recently become a subject of discussion for polyatomic reagents. Here again the thermal reaction can be observed, but subsequent thermal motion of the products is considered likely to have obscured the initial pattern of reaction. McCarty and Weiss studied the thermal dissociation of *p*-diiodobenzene by STM on Cu(111) at 77 K.[58] The two carbon–iodine bonds were observed to cleave within a short time of one another to yield pairs of I atoms approximately 4 Å apart. As this is significantly less than the I···I separation in the parent molecule (7.2 Å), the authors proposed that, rather than this being the initial product distribution, the I atoms had subsequently migrated under the influence of an attractive surface-mediated force.

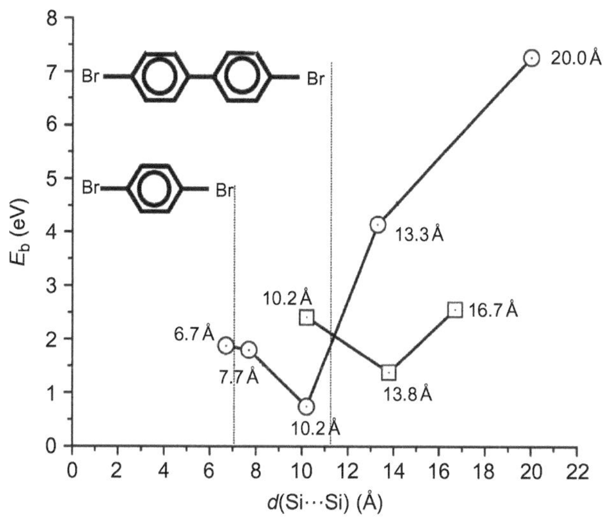

FIGURE 4.13 Barrier heights for differing dangling-bond separations selected to correspond to the adatom–adatom separations on the Si(111)-7 × 7 surface. The circles are for 1,4-dibromobenzene and the squares refer to 4,4-dibromobiphenyl. The dotted lines indicate the approximate Br–Br internuclear separations in these two molecules. Reproduced with permission from Ref. 20.

FIGURE 4.14 STM images of Pt(111), recorded after dissociative adsorption of O2 at around 160 K; black dots are oxygen atoms. 1.2 L O_2, 154 K, (V_{surf} = 0.13 V, I_{tun} = 0.8 nA 110 Å × 92 Å); the distances between the O atoms in the pairs are indicated in units of the lattice constant of length 2.775 Å. (Reprinted figure with permission from Wintterlin *et al.*[54] Copyright (1996) by the American Physical Society).

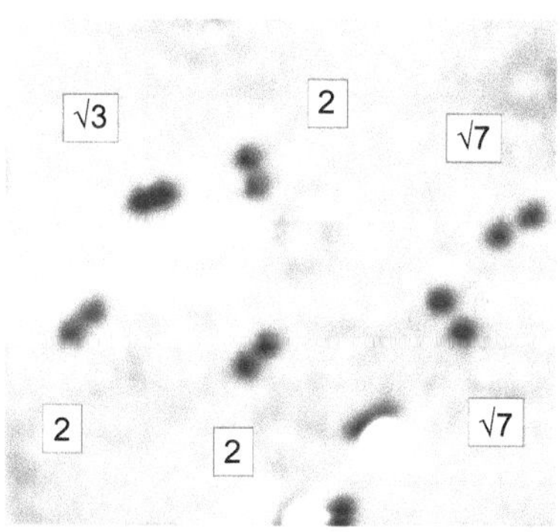

In other work, studies of the electron-induced reaction of benzene at metal surfaces were made on both Cu(001) at 4 K[59] and Cu(110) at 8 K.[60] Both studies concluded that an initially horizontal benzene molecule switched to a vertical geometry in an anionic state and transferred an H atom to the underlying Cu. The location of the H in the neighbourhood of the parent benzene was not reported.

In a further study,[61] the electron-induced reaction of iodobenzene to form phenyl and iodine at a 5 K Cu(110) surface was used as a means to establish the structure of linear clusters of this molecule, being observed for the first time.

More recent studies of the electron-induced reaction of *p*-diiodobenzene at Cu(110) held at 4.7 K yielded probability distributions for both distance, *P*(*d*),

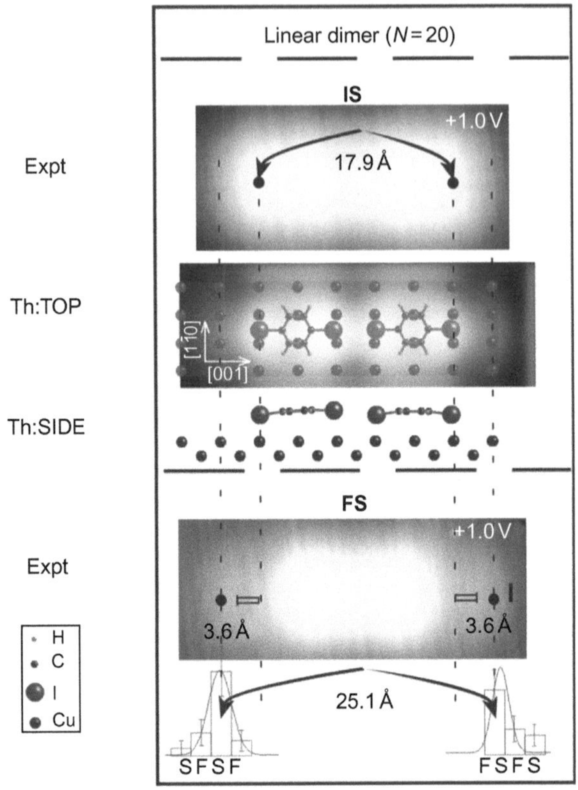

FIGURE 4.15 Electron-induced reaction of a linear dimer of *p*-diiodobenzene (pDIB) at Cu (110) at 4.7 K. In the initial, physisorbed state (IS), pDIB dimer is imaged ($V_s = +1$ V, $I_{tun} = 0.5$ nA) as a bright oval, whose essential features are well reproduced by theory as shown below (Th:TOP, Th:SIDE). The separation between the centres of the Iodine atoms is 17.9 Å. Following electron-induced reaction ($V_s = +1.3$ V), the FS consists of two Iodine atoms attached to the copper, with an observed mean separation of 25.1 Å. On average, the iodine atoms were displaced by only 3.6 Å from their initial positions. This is evidence of localized atomic reaction at a metal surface.[62] (A colour version of this figure is available in the Web version of this chapter.)

and angle, $P(\Theta)$, of the I atom reaction product, that is, the degree of localization of the atomic transfer reaction.[62] A high degree of localization of the product ('LAR' in the section heading 2.1) was observed, as evidenced in Figure 4.15. The observation of localized reaction in clusters of varied length, even at a metal surface, suggests that 'LAR' can provide a basis for 'molecular callipers' that imprint patterns with selected interatomic separation and alignment.

3. SELF-ASSEMBLY FOLLOWED BY PATTERN IMPRINTING

The formation of molecular patterns at surfaces can be achieved by numerous ingenious procedures involving the sequential processes of writing and etching.[63] Attempts are currently being made to multiplex this sequential approach by concurrently using many pens.[64] It is recognized, however, that this will not address the problem of scaling patterns up to the level of billions of atoms required for the creation of useful structures. For this, it will be necessary to multiplex the patterning procedure so that an entire patterned assembly is imprinted at one time. The first step in multiplexing is to assemble an entire physisorbed pattern in a single operation.

This multiplexed first step can conveniently be achieved by self-assembly; one shakes the molecular 'bricks' randomly by heat and the entire molecular-scale house assembles. Research into the self-assembly of varied patterns at surfaces is well advanced in many laboratories. The requirement of mobility for self assembly is, however, at variance with the requirement of durability for useable devices; there is little value in a house that collapses when you enter it. Nothing could be more calculated to shake loose a weakly attached structure than the envisaged passage of electric charge through molecular circuitry, one electron at a time, resulting in repeated cycles of attraction to the surface followed by repulsion.[65,66]

The physisorbed pattern is, therefore, in general only the first step in multiplexed pattern imprinting. The second step proposed as a means to the desired MSI[1,18,67] is more stable chemical attachment, in the place of physisorption. In order that the physisorbed pattern not be destroyed in the transition from physisorption to chemisorption, it is necessary that reaction be 'localized' to within an atomic spacing. Only then will the molecular 'printing press' print true.

It appears likely that this requirement for localization of reaction product to the near neighbourhood of the reagent molecule will often be met. The reason for this, it has been argued[9] (see above) is that the TSs of chemical reactions are stabilized, hence barriers to reaction reduced, by the concurrent existence of electronic charge shielding from one another the nuclei of the chemical bond being broken and at the same time further charge shielding the nuclei of the bond being formed, that is to say stabilization by the simultaneous existence of the 'old' and 'new' bonds. However, these concurrent partial bonds can only coexist if adjacent. For this reason, 'localized' reaction seems likely to be the preferred reaction pathway, particularly at energies close to the threshold energy for dissociative attachment.

3.1. Imprinting Circles

Physisorbed circles of benzene molecules were first reported by Brown *et al.* for benzene[68] on Si(111) at 78 K and were attributed to charge-donation binding to the electrophilic 'middle' atoms of this surface. These middle atoms constitute circles of 12 raised 'adatoms' traversing six adjacent half-unit cells.

For a test of MSI, the physisorbed species was methyl bromide, a surface-halogenation agent.[18] A limitless field of unreacted physisorbed CH_3Br

molecular circles on Si(111)-7 × 7, comprising 12 CH_3Br molecules per circle, was obtained at 50 K surface temperature (see Figure 4.16B).

In a first study of MSI, these physisorbed molecular circles were shown to 'imprint' at the surface as chemisorbed Br, without detectable alteration in their pattern, when irradiated with light (193-nm excimer radiation) or electrons (randomly distributed charge from the STM tip, at a bias of $V_{surf} = +2.5$ V). In subsequent work on the same system, these observations were extended to the case of MSI induced by thermal reaction at approximately 80 K; Figure 4.16 shows a bright circle of physisorbed CH_3Br (B and C) at the left, fully reacted to form a similar circle of dark chemisorbed Br at the right (D and E by photolysis, F by electron-induced reaction). The conversion from physisorbed molecules to chemisorbed reaction product is rendered unambiguous[18] by the fact that the physisorbed pattern is stable only up to approximately 50 K, whereas the chemisorbed one is stable to approximately 500 K. Though the circular pattern over 12 surface 'middle' adatoms was unchanged in general appearance, a substantial chemical change had taken place. All CH_3 radicals left the surface and the residual Br, being an order of magnitude more strongly bound than the physisorbed precursor, had moved vertically much closer to the surface and shifted by on the order of 1 Å laterally. A visualization of this MSI event is in Figure 4.17. The molecular dynamics, obtained from an *ab initio* calculation,[17] can be seen in Figure 4.3, illustrative of the LAR underlying MSI.

Closer examination of the physisorbed state topographs for CH_3Br[17] revealed evidenced of four geometries, two at 50 K and two at 80 K, of which the most strongly displaced at 80 K had shifted approximately 1 Å away from the centre of the underlying Si atom. STM height profiles computed by DFT showed similar shifts in the simulated image peak as the physisorbed CH_3Br approached the crest of the calculated 0.32-eV barrier to reaction. The multiplicity of moderately shifted physisorbed states, with some metastable configurations at higher energy in the region of the TS, offers unrealized possibilities for future experimentation. One can, for example, envisage experiments in which the dynamics of electron-induced reaction is explored by projecting the reagent onto the charged intermediate state from differing metastable starting points, thereby probing this little-understood intermediate, and achieving a means to controlling the reaction outcome.

Of the three modes of bringing about MSI, photolytic, electron-induced and thermal, the last, at the present time, offers the best possibility for theory since the entire event is on the ground-state PES. In Figure 4.3, the IS, TS and final state are shown for characteristic nudged elastic band (NEB) calculations performed for the thermal bromination of Si(111) by CH_3Br.[17] As noted above, the localization of the reaction to the silicon atom beneath the Br-end of the reagent molecule was clearly evident. The charge density isosurface contours shown are consistent with the coexistence in the TS of partial bonding in the extended C—Br (the bond being broken) and Br—Si (the bond

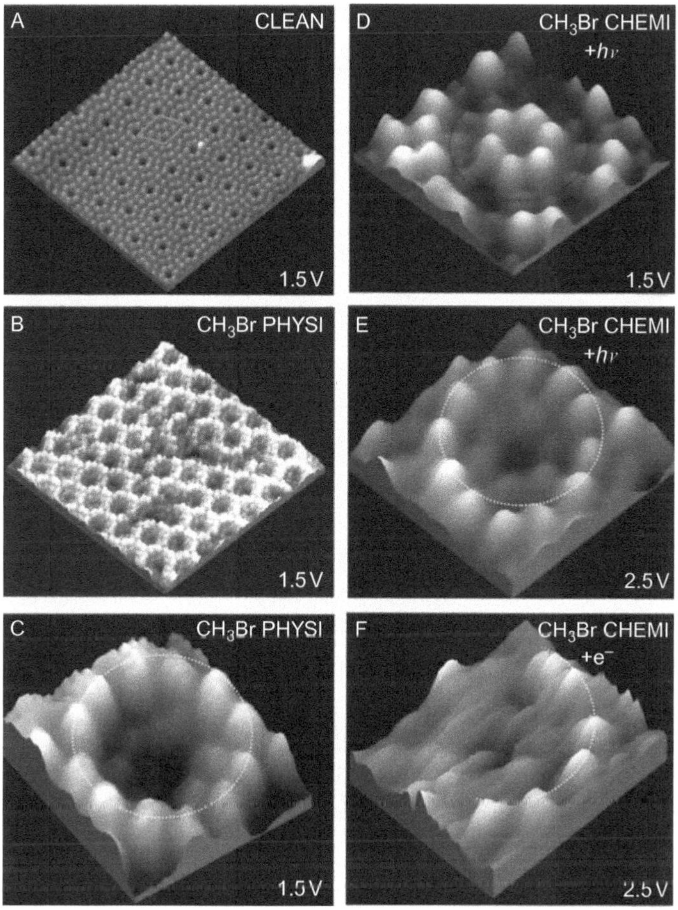

FIGURE 4.16 (A) STM image of the clean Si(111)-7 × 7 surface at 50 K. A 7 × 7 unit cell is indicated. V_{surf} = 1.5 V, I_{tun} = 0.2 nA, ~200 Å × 200 Å. (B) STM image of physisorbed CH_3Br (ad) on the 50 K Si(111)-7 × 7 surface at a coverage of 0.41 monolayer. Physisorbed molecules appear as protrusions over the middle adatoms. V_{surf} = 1.5 V, I_{tun} = 0.2 nA, ~200 Å × 200 Å. (C) Zoomed-in STM image of a single ring of physisorbed CH_3Br on Si(111) surface (indicated by the dotted circle), as in (B) but ~30 Å × 30 Å. (D) Chemisorbed Br on Si(111) surface after photolysis of (three successive applications of) physisorbed CH_3Br(ad) at 50 K. Br (beneath dotted circle) appears as depressions on the middle adatoms. V_{surf} = 1.5 V, I_{tun} = 0.2 nA, ~30 Å × 30 Å. (E) STM image of chemisorbed Br imprints on the middle adatoms (indicated by a dotted circle) as in (D) but with V_{surf} = 2.5 V. (F) STM image of chemisorbed Br from electron-induced reaction at the middle adatoms (dotted-in) obtained by scanning (a single application of) physisorbed CH_3Br (ad) at 2.5 V (scans from lower left to upper right); V_{surf} = 2.5 V, I_{tun} = 0.2 nA, ~30 Å × 30 Å. Adapted with permission from Ref. 18. (A colour version of this figure is available in the Web version of this chapter.)

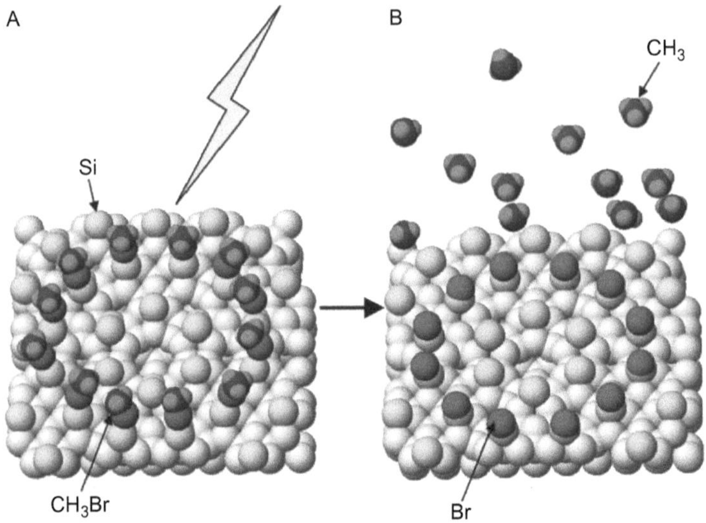

FIGURE 4.17 Schematic representation of (A) physisorption of CH_3Br on Si(111) surface with Br pointing down, and (B) chemisorbed Br on middle adatom positions, after photolysis or electron-impact at 50 K. Reproduced, with permission, from Ref. 18. (A colour version of this figure is available in the Web version of this chapter.)

being formed). LAR was previously ascribed to this preference for reaction to occur by way of a 'concerted' mechanism, with old and new bonds coexistent in time and therefore, necessarily, in space.[9]

A further feature of this CH_3Br reaction with this particular surface, Si (111), evident in the reaction path is the progression from a horizontal IS (C—Br parallel to the surface) to a vertical final state. This requirement is in conformity with the experimental observation that the CH_3 radical was never found attached to the Si(111) surface, recoiling instead into the gas phase.

The horizontal to vertical transition pictured in Figure 4.3 above is not a general phenomenon. The thermal reaction of CH_3Br with a different face of silicon, Si(100), was shown (see Figure 4.6, Figure 4.7) to imprint three different patterns at the surface, all of which share the characteristic that the horizontal ISs (see Figure 4.7) lead to 'horizontal' outcomes in which *both* ends of the reagent molecule remain attached (as chemisorbed CH_3 and Br) to the surface. Generalization of these profound differences is not possible as yet. By analogy with the gas phase, one might suppose that for CH_3Br/Si(111) the PES is 'repulsive' with a major part of the reaction energy being channelled into Br—CH_3 repulsion, whereas for CH_3Br/Si(100), the interaction is more 'attractive' with reaction energy preferentially channelled into Br—Si vibration, thereby permitting CH_3 to remain at the surface.

3.2. Forming Lines

There exist a number of proven methods for forming lines at the commercially important Si(100) surface. Group II, III and IV metals[69–75] bismuth[76–81] and rare earths[82–91] reacted on the heated surface and formed atomic lines perpendicular to the Si dimer rows. Chemical chain reactions across the hydrogenated surface were shown to attach molecular reagents in lines running along or perpendicular to the surface's dimer rows.[92–96]

The interest in the context of this chapter is in the application of the MSI approach to linear patterning by atoms and molecules. The requirements are first that a reagent molecule be found that is capable of self-assembly into physisorbed lines across the rough surface of silicon, and second that a means be found to induce localized chemical reaction of molecules or their fragments from such physisorbed lines.

It has been found[97] that haloalkane molecules of low molecular weight are mobile on Si(100)-2 × 1 at room temperature, self-assembling into lines by a process in which the physisorbed adsorbate on one dimer row causes charge transfer, and hence buckling at an adjoining dimer row, with resultant chain growth of physisorbed molecules. This growth process is unidirectional and takes its direction from the dipole moment in the adsorbate (indicated by the black dipole arrows (extending from plus to minus, with a calculated dipole moment of 4.9 Debye shown in Figure 4.18, top panel). As this surface dipole is crucial in buckling the adjacent dimer pair and thereby propagating the physisorbed chain, the mechanism of chain growth was termed dipole-directed assembly (DDA).[97] For dichloropentane, chains of single molecules were formed with their two Cl atoms over Si atoms, as shown in Figure 4.18, top panel. The halogen atoms in the adsorbate molecule were shown to donate charge as illustrated in Figure 4.18 (bottom panel). This shows the source of binding by charge donation from the lone pairs of Cl atoms of physisorbed molecules into surface dangling bonds (hence loss of charge from Cl to the underlying Si and gain of a corresponding charge at the other Si of that silicon dimer pair). This charge donation from the physisorbed halogen atom conforms with earlier descriptions of the binding of CH_3Br at electrophilic Si middle atoms of Si(111) to form physisorbed circles, given previously in this section.[17]

3.2. Imprinting Lines

The MSI of lines, rather than circles as illustrated above, has recently been demonstrated by linear self-assembly of molecular reagent, followed by chemical reaction. Energization of the physisorbed reagent to induce reaction was achieved variously by heat, electrons or light, resulting in every case in the 'localized reaction' that is the prerequisite for imprinting without loss of the prior physisorbed pattern.

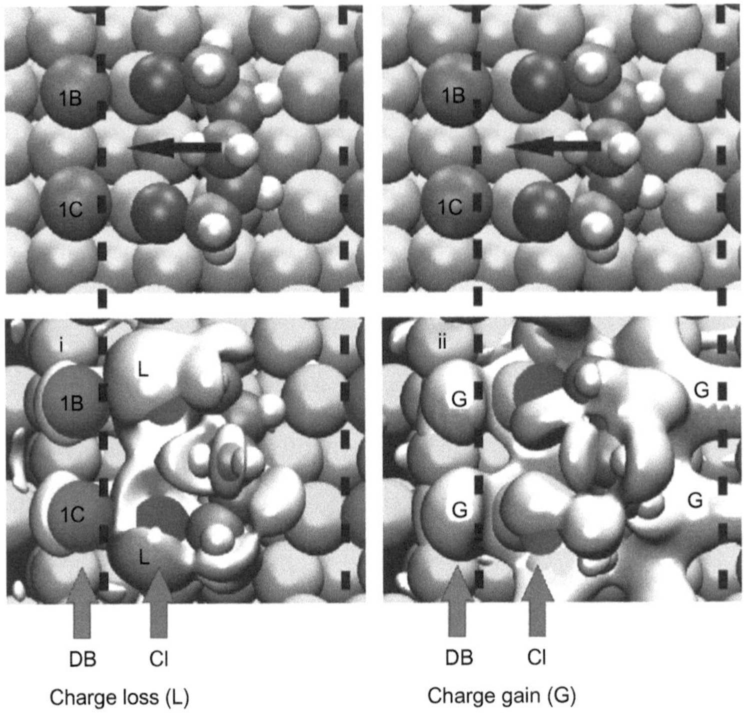

FIGURE 4.18 Mechanism of formation of the molecular lines of 1,5-dichloropentane (theory). *Top*: schematic (twice repeated for convenience) showing the physisorption configuration of 1,5-dichloropentane at Si(100)-2 × 1 along with black dipole-arrows (see text). *Bottom*: charge redistribution due to physisorption a dichloropentane molecule atop two silicon dimers. *Bottom left*: Loss (L) in electronic density within the white contour shown (−0.004 e/Å ^{3}). *Bottom right*: gain (G) in electronic density within the white contour (+0.004 e/Å ^{3}). The red arrows marked Cl indicate the approximate locations of the halogen atoms that lose negative charge (sky blue). The green arrows marked DB indicate the locations of the Si dangling bonds that gain charge (shown in red). The vertical black dashed lines denote the centres of the Si dimer rows. Adapted with permission from Ref. 97. (For interpretation of the references to colour in this figure legend, the reader is referred to the Web version of this chapter.)

MSI of lines was demonstrated for 1-monochloropentane (CP) self-assembled by DDA as molecular *pairs*. Figure 4.19[49] shows a number of such self-assembled lines of CP pairs, each terminating in a buckled growing point, labelled 'b' in Figure 4.19A′. These physisorbed lines were used as a starting point for the imprinting of chemisorbed lines of atomic Cl, following energization by each of the three modes: electrons, A → B; heat, C → D; or photons, E → F.

Figure 4.19 (top) shows the effect of three whole-area STM scans at a surface bias of +2.6 V, current of 0.2 nA; this generalized application of electrons caused the physisorbed molecular line to convert into a dark line, identified by scanning spectroscopy as comprising pairs of chemisorbed Cl atoms. (Section 4.2 identifies the dynamics of halogen pair formation as

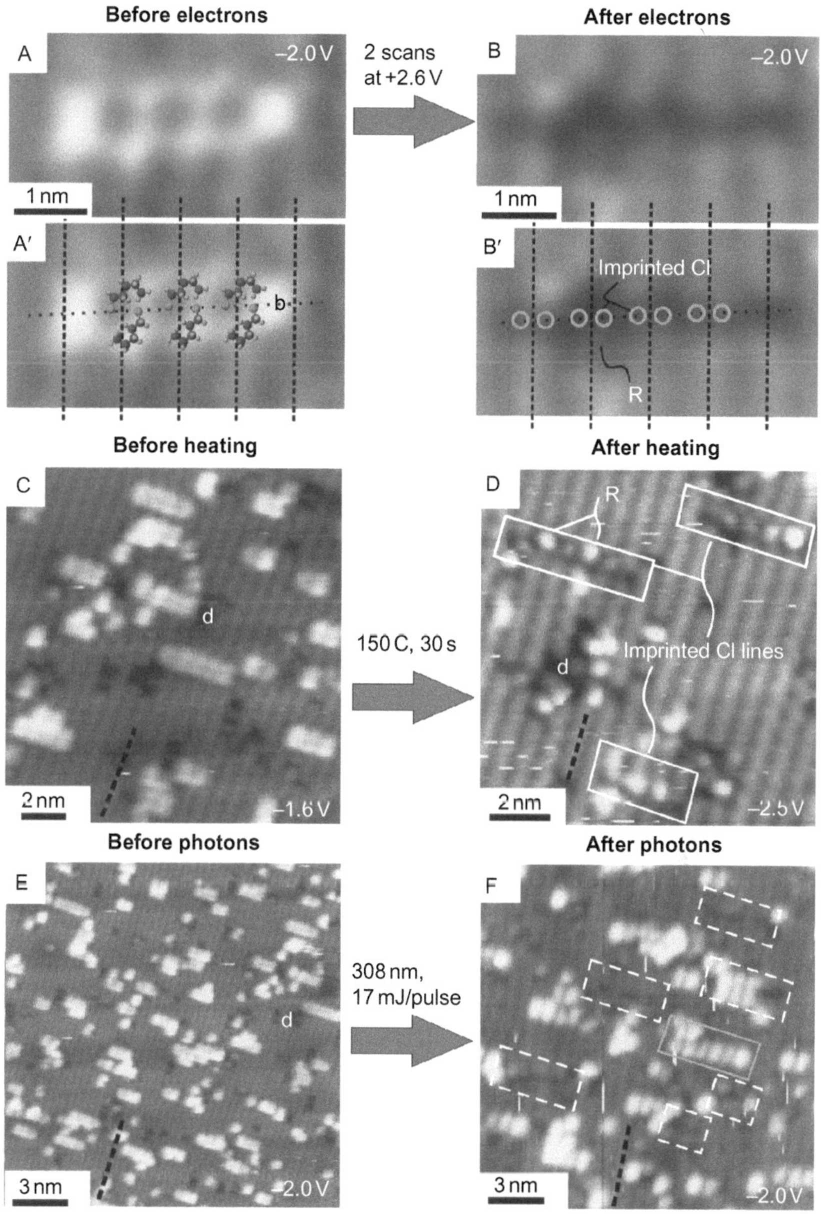

FIGURE 4.19 Three modes of molecular-scale imprinting of lines on Si(100)-2 × 1, separately achieved using electrons (A,A′ → B,B′), heat (C → D) or photons (E → F). (Image scales and surface bias voltage shown on each image. All images recorded with $I_{tun} = 0.2$ nA.) Adapted with permission from Ref. 19. (A colour version of this figure is available in the Web version of this chapter.)

involving imprinting of the first Cl, followed by 'cooperative reaction' of the second.) In a minority of cases, the reaction products bound covalently to the surface included a pentyl radical, *R*, beside a chemisorbed Cl; this was not, however, a requirement for the surface chlorination reaction to occur (see Section 4.2). Figure 4.19 (middle) shows a similar reactive outcome, namely, dark lines of chemisorbed Cl pairs, due to the application of heat (150 °C for 30 s). Reaction (once again involving 'Cooperative' dynamics at each Si dimer) was highly localized. The third pair of panels (Figure 4.19, bottom) shows the effect of 2 h of laser irradiation at 308 nm (17 mJ/pulse, 5 Hz); localized photoreaction again gave rise to dark lines of Cl pairs. All three types of energization resulted in MSI; MSI of a chemisorbed line of Cl atoms from a physisorbed line of physisorbed reagent chloropentane molecules.

4. MODES OF REACTION

This section describes special modes of surface reaction, additional to the normal atomic transfer and dissociative attachment described above, found recently by the application of STM to single-molecule dynamics.

4.1. Direct and Indirect Reaction

In work on 1-halododecane reaction at a Si(111)-7 × 7 surface, the alternative reaction pathways were termed, respectively, 'daughter' and 'parent' mediated.[98] They are renamed here, less ambiguously, as 'direct' and 'indirect' reaction pathways. The long-chain (12 carbon atom, dodecane) halides were observed by STM at low temperature standing vertically on the halogen end, and at room temperature lying horizontally on the surface. As the reactive entity, Br, is itself responsible for holding the vertical dodecane in place at the surface, we term the reaction 'direct'. (Previously, we called it 'daughter-mediated', as the 'daughter' Br product also mediated the attachment). For 1-bromododecane, this 'direct' reaction was found to take place over an energy barrier of 5 kcal/mol (0.2 eV) in the measured temperature range from −60 to +60 °C. 'Direct' reaction gave rise to chemisorbed Br in the absence of accompanying reacted dodecyl radicals, *R*. This was understandable, as the R-end of the vertical dodecane was not physisorbed in the vertical reagent configuration responsible for direct reaction. To summarize, 'direct' reaction was thought to result from encounters between vertical bromododecane molecules, with only the reactive specie, Br, in contact with the substrate, first in the physisorbed reagent and then in the chemisorbed product.

At a higher temperature (75 C), horizontal bromododecane could, for the first time, be observed to convert at a measurable rate (~1 h) into chemisorbed Br, accompanied in this case sometimes by an adjacent chemisorbed radical, *R*.

An Arrhenius plot led to an activation energy for this 'indirect' reaction path of 28 kcal/mol (1.2 eV). The reaction is termed 'indirect' due to the prior existence of a precursor (or 'parent') molecule physisorbed at the surface through interactions other than Br· · ·Si, that is, bound by other than the reacting moiety. For the horizontal dodecane, the attachment to the surface in the reagent is by way of R· · ·Si attraction. The reason that the distinction is significant is that in 'indirect' reaction, the physisorbed reagent must be distorted to bring it into the reactive TS configuration, with the reagent atom (Br in the present instance) at its bonding distance from the surface.

Accordingly, the large observed difference between activation barriers for 'direct' and 'indirect' pathways was attributed to the fact that in the former case (low Ea, 0.2 eV), involving the vertical molecule, the reactive Br was free to seek out the minimum energy path, whereas in the latter (high Ea, 1.2 eV), the Br was severely constrained in its motion by the physisorbed hydrocarbon tail. Direct reaction over a low energy barrier and indirect (precursor-mediated) reaction over a higher barrier, for a single reagent reacting at a given surface, is anticipated to be a general phenomenon.

4.2. Cooperative Reaction

In 'cooperative reaction', attachment to one atom of a silicon surface causes a second reaction to take place with unit probability at an adjacent atom.[99] 'Cooperativity' has previously been invoked to describe adsorbate–adsorbate interactions at elevated coverages.[100–103] The present usage is different, referring to action at a distance operating by charge transfer through the substrate.

Specifically, addition of a fluorine atom or a chlorine atom to one side of a Si dimer pair of Si(100)-2 $\times$ 1 in the presence of further adsorbate was found invariably to induce halogenation of the other atom of the pair. The reactants were physisorbed 1-fluoropentane or physisorbed 1-chloropentane. Cooperative reaction was induced either thermally or by a single-electron process.

For the case of thermal fluorination, *ab initio* calculations gave a barrier of 1.4 eV for the first atom transfer (see Figure 4.20; experiment gave 1.2 eV) and a barrierless reaction for the subsequent second atom transfer, the 'cooperative' event which was inferred to occur some femtoseconds later. These sequential events cannot be time resolved by STM which has a response time of, at best, some milliseconds. The computed events suggest that the formation of the first silicon–halide bond gave rise, by charge flow through the substrate, promptly to a reactive 'dangling bond' at the adjacent Si. The process can be pictured as the opening of the double bond that previously held together Si=Si with, sequentially, one fluorine attaching to each silicon atom.

In the following section, it will be shown that a succession of such events, clearly detectable by STM, can propagate a chain reaction.[99,105]

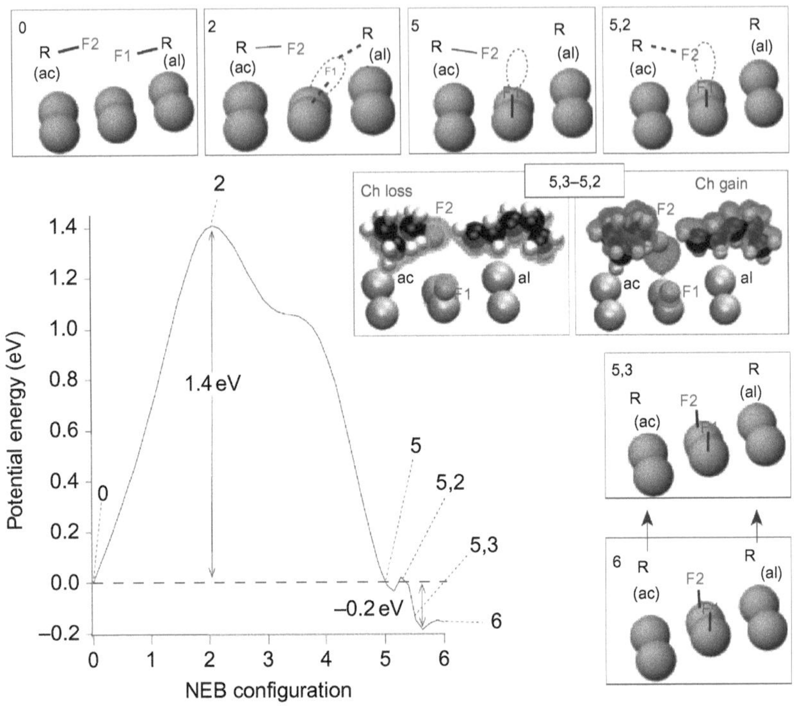

FIGURE 4.20 Calculated minimum-energy path for thermal cooperative bifluorination. Twelve NEB configurations were calculated (0–6 and 5.1–5.5); only two of the five extra configurations calculated between 5 and 6 are indicated. The barrier to the reaction was calculated as 1.4 eV, in satisfactory agreement with the experimentally determined 1.2 eV. Insets show how the thermal reaction proceeds. (A movie of the reaction dynamics is available online.[104]) The silicon dangling bond on the silicon atom adjacent to F1 is indicated as an ellipse (dotted line). The first fluorine atom is transferred fully by stage 5. The second fluorine atom, F2, then transfers rapidly, between stages 5.2 and 5.3. Transfer of F2 is shown in the red inset to involve electron charge (Ch) loss from the region of C–F2, and in the blue inset to involve electron charge gain between F2 and the Si to which it will bond. Gain and loss contours were obtained by subtracting the charge density for configuration 5.2 from that for configuration 5.3. The reaction is essentially complete by stage 6, at which both fluorines are attached fully to the surface, and both radicals have moved away. Reproduced with permission from Ref. 99. (A colour version of this figure is available in the Web version of this chapter.)

4.3. Chain Reaction

Chain reaction at surfaces has been a lively area of study in recent years, in part because of the prospect of fabricating nanowires.[92,106–110]

Cooperative reaction based on charge transfer through the surface of Si (100), described in the previous section, suggests a further mechanism to propagate reaction. For extended chains, it requires that a means be found to bridge the gap between dimer rows, as cooperative reaction described in the previous section can be relied on to bridge dimer pairs within each row.

This 'missing link' is provided by the dissociative attachment of methyl chloride which occurs IR, thereby chemically attaching CH_3 to the inside Si of one dimer row and Cl to the inside Si of the next adjacent dimer row.[105] These chemisorbed species give rise, by surface-mediated charge transfer, to reactive dangling bonds at the outer Si atoms of the dimer pairs at each of the two silicons, one adjacent to CH_3—Si and one adjacent to Si—Cl. These dangling bonds (in effect, free radicals) in turn react—in the cooperative mode described above, as they constitute the second atom of a dimer pair—to dissociate further incoming CH_3Cl molecules, thereby constituting the growing points for chains of indefinite length. Figure 4.21[105] shows schematically the dominant mode of chain growth. Figure 4.22 shows an STM image of a growing chemisorbed chain, with the observed characteristic CH_3 and Cl alternation. Figure 4.23 shows an *ab initio* calculation of the charge accumulation at the two dangling bonds beside a chemisorbed CH_3 and Cl, the greater charge gain being beside the Cl, found, accordingly, to be the preferred point of attachment (in experiment and theory) for the electrophilic CH_3 end of the incoming $CH_3Cl(g)$ molecule which, on dissociation, will provide the next link in the chain.

4.4. Recoil Reaction

In the preceding text, we stressed the propensity for reactions to occur locally. This provides a basis for the 'imprinting' of self-assembled patterns. As, however, reactions that occur at normal temperatures over low energy barriers are for the most part exothermic, localization of reaction depends upon the efficient channelling of reaction energy into the substrate. In electron-induced reaction or photoelectron-induced reaction by way of a short-lived ionic state, it has been suggested that

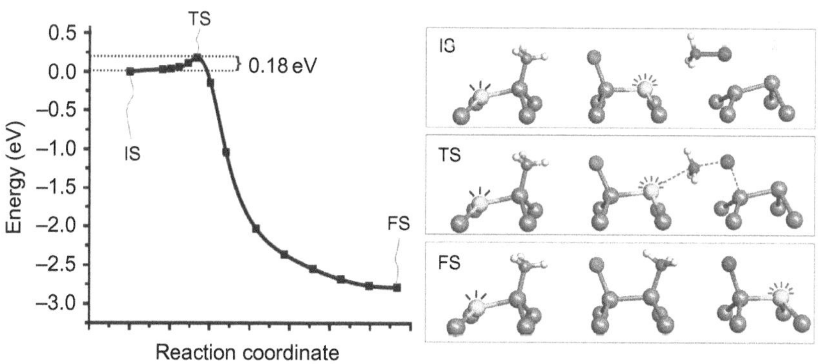

FIGURE 4.21 Observed dynamics of DA chain growth. The schematic summarizes the modes of surface-mediated DA chain reaction of CH_3Cl on Si(100)-2 × 1. The CH_3 product of an incoming CH_3Cl reacts with the bright DB end of the existing chain, adjacent to a Cl atom, forming a new bright DB adjacent to a newly attached Cl. The initial state (IS), transition state (TS) and final state (FS) pictured at the right are labelled on the corresponding energy profile at the left. Adapted with permission from Ref. 105. (A colour version of this figure is available in the Web version of this chapter.)

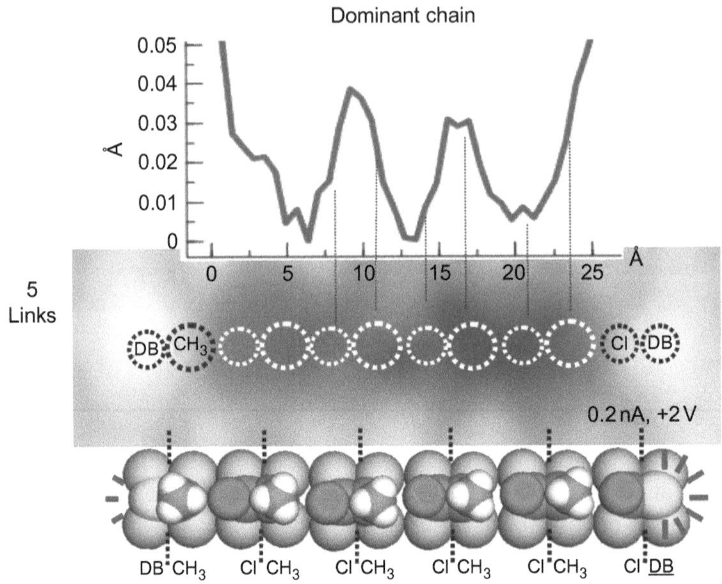

FIGURE 4.22 Dissociative attachment and chain growth. A height profile of the five-link chain shows the alternation of the CH_3 and Cl groups in mid-chain. Adapted with permission from Ref. 105. (A colour version of this figure is available in the Web version of this chapter.)

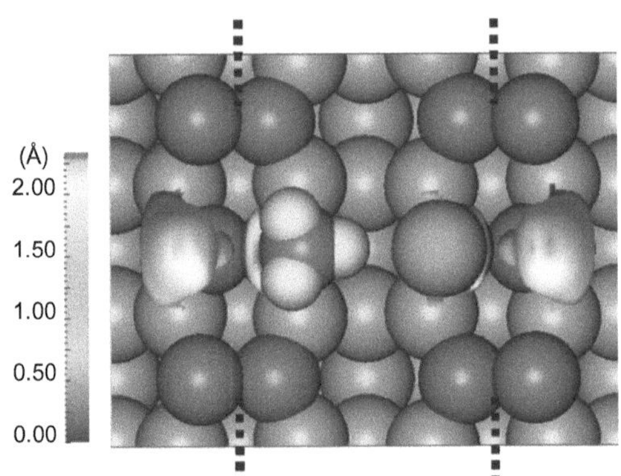

FIGURE 4.23 *Ab initio* findings for DA chain growth. Computed electron gain at a dangling bond to the left of CH_3 (grey and white) at one Si dimer pair, and the slightly larger gain at a dangling bond to the right of Cl (green) at the adjacent dimer pair. The colours indicate the relative height of the electron-density isosurface with 0.005 e/Å^3. The dominant mode of chain growth is by attachment of the next (electrophilic) CH_3 at the dangling bond adjacent to the Cl. Adapted with permission from Ref. 105. (For interpretation of the references to colour in this figure legend, the reader is referred to the Web version of this chapter.)

the substantial excess energy is transferred to the substrate by the departing electron.[18] This could explain the persistence of localized reaction even when induced by high-energy photons (193 nm = 6.4 eV).

By contrast, recently, examples have been reported, in which highly exothermic surface reaction at Si(100) evidenced sufficient torque in a reaction product to cause long-range rotational recoil.[111] The recoil, it is important to note, was of a physisorbed specie. The recoiling moiety was not a free radical (such as a halogen atom) which would be expected to interact strongly with surface atoms, but a molecule. The recoil was over distances as great as 200 Å.

In each of the six reported cases, the recoiling molecule has been ethylene or a derivative of ethylene. The significance of the ethylenic bond is likely to be that in all these cases, there is substantial energy release (exoergicity) that propels the molecule across the surface, stemming from the formation of a C=C double bond. The observed distance of recoil was an average of 29 Å at a room temperature surface and 87 Å at a 150 °C surface. This is characterized as 'long-range recoil' in contrast to the hopping diffusion of less than 5 Å reported by previous workers in a variety of contexts.[112–120]

Three of the examples of long-range product recoil involved thermal dihalogenation of a silicon surface by a 1,2-dihaloethane (XCH_2CH_2X, X=F, Cl or Br) with several electron volts of energy release, to give recoiling ethylene. Three other examples involved the electron-induced reaction of chemisorbed alkenes: ethylene, propene and trans-2-butene. These were initially and again finally following long-range recoil, disigma bound to the surface.

Figure 4.24 illustrates one case of the electron-induced recoil of ethylene; the ethylene surmounts an obstacle without deflection, as if rolling rather than sliding.

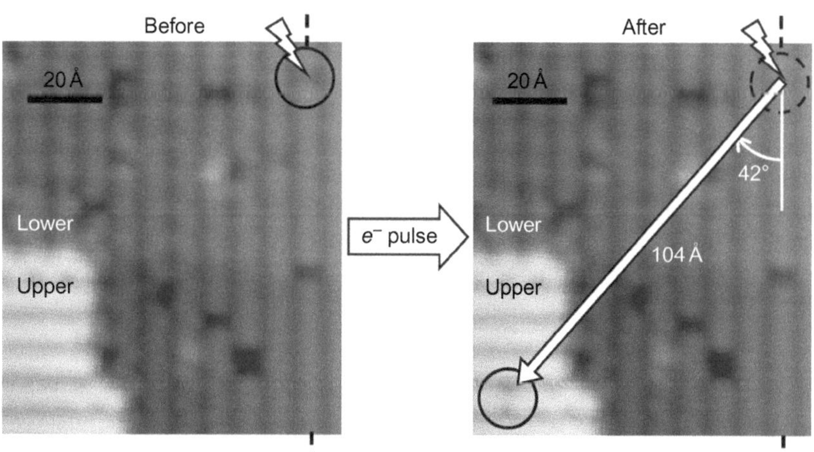

FIGURE 4.24 Electron-induced ethylene migration. STM images ($V_{surf}=-1.7$ V, $I_{tun}=0.1$ nA, 25 °C) of identical areas (left) before and (right) after an electron pulse ($V_{surf}=-3.5$ V, $I_{tun}=0.1$ nA, 2 s) on a chemisorbed OD ethylene. The white arrow shows ethylene migration by 104 Å passing from a lower onto an upper terrace. Adapted with permission from Ref. 123. (A colour version of this figure is available in the Web version of this chapter.)

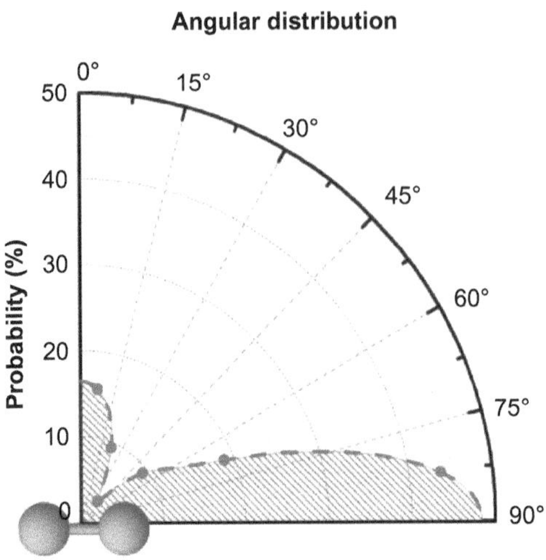

FIGURE 4.25 Angular distributions of electron-induced ethylene migration at a surface temperature of 150 °C (solid red). The blue dumbbell gives the direction of the C—C axis of ethylene chemisorbed in the on dimer (OD) initial state applicable to these electron-induced recoil events. Adapted with permission from Ref. 123. (For interpretation of the references to colour in this figure legend, the reader is referred to the Web version of this chapter.)

Figure 4.25 shows the preferred direction of the electron-induced recoil. The marked directional dependence evident at the hot surface is a strong indication that the process is indeed recoil, and not thermal diffusion which would be expected to favour motion *along* the dimer rows, the converse of what is observed.

Figure 4.26 shows *ab initio* evidence of torque for both ethylene recoiling from thermal reaction of dihaloethane at Si(100), and also ethylene recoiling from electron-induced disigma bond breaking at Si(100).

In all these cases, the recoiling ethylenic molecule traversed, for substantial distances, the rough surface of silicon known to be resistant to molecular migration. (It is this roughness that makes possible the observation of stationary physisorbed molecules under an STM tip at room temperature on semiconductors, in marked contrast to metals.) The recoiling ethylenic molecule evidenced a preferred direction of motion along the C—C bond axis of its initial OD chemisorbed state, recoiling, therefore, perpendicular to the dimer rows of the surface. The recoil takes the physisorbed ethylene undeflected across raised obstacles, such as upward steps in the surface. This is evidence against translation across the surface but is suggestive of rotation. More direct evidence of rotation comes from the observation that recoiling propene suffers end-to-end inversion in half of the recorded events. Rotational recoil is also supported by *ab initio* calculations that show, for both the thermally induced and the electron-induced reaction, asymmetric forces acting on the carbon atoms at either end of the ethylenic bond being formed. The observation of exothermicity, directionality, persistence of motion and end-to-end inversion supports a picture of a rotationally hot recoiling molecule cartwheeling across the rough surface of Si(100), losing energy as it proceeds (the distance distribution, $P(d)$, exhibits exponential decay) until being ultimately trapped by reactive reformation of a pair of sigma bonds to the surface.

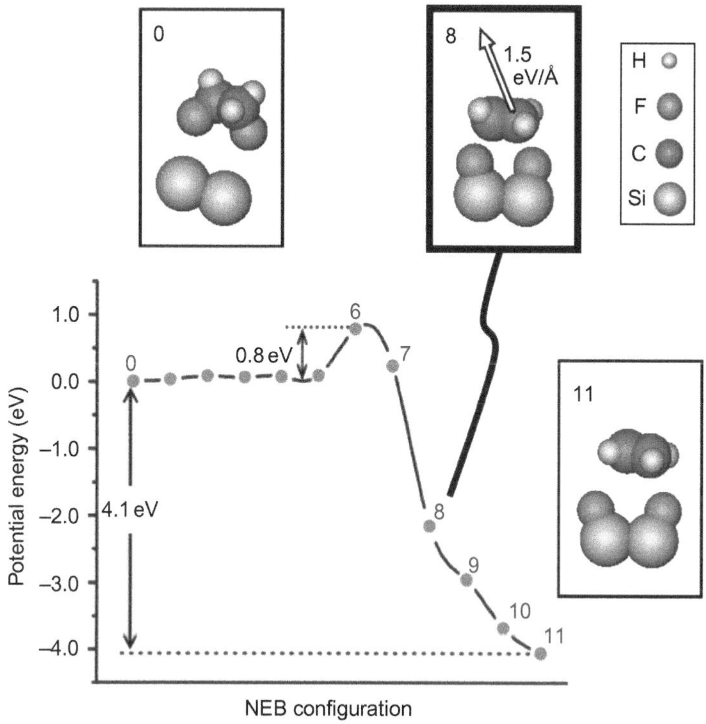

FIGURE 4.26 DFT calculations show recoil of ethylene as a product of thermal reaction. Calculated minimum-energy path for thermal dissociation of DFE on Si(100). Twelve NEB configurations were calculated that spanned the initial state, IS (step 0, see insert) to the final state, FS (step 11, see insert). A white arrow on step 8 shows differential forces on the carbon atoms inducing torque. Adapted with permission from Ref. 123. (A colour version of this figure is available in the Web version of this chapter.)

In much of the preceding discussion, we noted the prevalence of localized reactive events. Here, we have a rarer example of delocalized reaction stemming from the formation of a mobile physisorbed molecule. This molecule is formed by recoil in a 'repulsive' exothermic reaction.[121] Asymmetric repulsion applies torque, and the rotationally hot molecular product traverses the surface. As noted some time ago,[122] a highly rotationally excited molecule can be expected to have reduced reactivity due to the fact that it rotates out of the geometry favourable to reaction. This temporary suppression of reactivity makes possible the observed delocalized reaction.

5. CONCLUSION

In this chapter, on the chemisorptions imprinting of patterns at semiconductor and metal surfaces, we began by stressing the concept of the 'localization' of surface reaction to the close vicinity of the reagent molecule. This accounted for the frequent observation that a pattern existing either in a single reagent molecule or in a self-assembled group of reagent molecules survived to a significant extent when chemically 'imprinted' (chemisorbed) at the surface. The rationale offered for this localization of chemical reaction, on the basis of *ab initio* calculation, was that the energy of the TS and hence the activation energy for reaction were minimized by the concurrent existence of partial chemical bonds in the reagent and product locations. Tersely stated, chemical reactions prefer to occur in a 'concerted' fashion rather than a 'sequential' pattern of bond breaking and bond formation. This necessitates proximity of the 'old' and the 'new' chemical bonds—that is restriction of the reaction to a localized region.

Having made this point, we were humbled by nature, which offered a striking variant. The variant was for exoergic reactions that released their energy in the form of repulsion between a physisorbed (note 'physisorbed') reaction product and the surface. If that repulsion were directed perpendicular to the surface, one could expect desorption of the physisorbed product, rather than surface reaction. However, if the repulsion is directed along the surface, it can produce sufficient torque to induce cartwheeling rotational motion in a physisorbed product, suppressing reaction and transporting the physisorbed product.

As the physisorbed molecule rolls, it cools. When the rotational energy has been sufficiently reduced, it 'reacts', that is, it undergoes chemisorption at the surface. The chemisorptions event is still 'localized', but through prior rolling recoil in a physisorbed state, its locale has changed.

ACKNOWLEDGEMENTS

We are grateful to our coworkers for their generous permission to use figures in advance of publication: Profs. W. A. Hofer, H. Guo, W. Ji, Drs. L. Leung, T. B. Lim, K. R. Harikumar and A. Zabet-Khousousi. Financial support from the Natural Sciences and Engineering Research Council of Canada (NSERC), Photonics Research Ontario (PRO) an Ontario Centre of Excellence, the Xerox Research Centre Canada (XRCC) and the Canadian Institute for Advanced Research (CIFAR) for the research of this laboratory is gratefully acknowledged.

REFERENCES

1. McNab IR, Polanyi JC. Patterned atomic reaction at surfaces. *Chem Rev* 2006;**106**:4321.
2. Avouris Ph, Wolkow R. Atom-resolved surface chemistry studied by scanning tunneling microscopy and spectroscopy. *Phys Rev B* 1989;**39**:5091.
3. Takayanagi K, Tanishiro Y, Takahashi M, Takahashi S. Structure analysis of Si(1110-7 × 7 reconstructed surface by transmission electron diffraction. *Surf Sci* 1985;**164**:367.
4. Brommer KD, Galvan M, Pino Jr. AD, Joannopoulos JD. Theory of adsorption of atoms and molecules on Si(111)-7 × 7. *Surf Sci* 1994;**314**: p. 57.

5. Wang YL, Gao H-J, Guo HM, Liu HW, Batyrev IG, McMahon WE, Zhang SB. Tip size effect on the appearance of a STM image for complex surfaces: Theory versus experiment for Si(111)-(7 × 7). *Phys Rev B* 2004;**70**:073312.
6. Owen JHG. Competing interactions in molecular adsorption: NH_3 on Si(001). *J Phys Condens Matter* 2009;**21**:443001.
7. Cho IK, Kim YK, Yeom HW. Temperature-dependent adsorption and dissociation behaviors of NH_3 on Si(111)7 × 7: A high-resolution core-level photoemission study. *Phys Rev B* 2006;**73**:115328.
8. Kang MY. Theory of the site-selective reaction of NH_3 with Si(111)-(7 × 7). *Phys Rev B* 2003;**68**:205307.
9. Lu PH, Polanyi JC, Rogers D. Electron-induced "localized atomic reaction" (LAR): Chlorobenzene adsorbed on Si(111)-7 × 7. *J Chem Phys* 1999;**111**:9905.
10. Naumkin FY, Polanyi JC, Rogers D, Fisher A, Hofer WA. Electron induced attachment of chlorinated benzenes to Si(100)2 × 1. *Surf Sci* 2003;**547**:324.
11. Wolkow RA, Lopinski GP, Moffatt DJ. Resolving organic molecule–silicon scanning tunneling microscopy features with molecular orbital methods. *Surf Sci* 1998;**416**:L1107.
12. Cao Y, Deng JF, Xu GQ. Stereo-selective binding of chlorobenzene on Si(111)-7 × 7. *J Chem Phys* 2000;**112**:4759.
13. Wolkow RA. Controlled molecular adsorption on silicon: laying a foundation for molecular devices. *Annu Rev Phys Chem* 1999;**50**:413.
14. Jiang G, Polanyi JC, Rogers D. Electron and photon irradiation of benzene and chlorobenzene on Si(111)7 × 7. *Surf Sci* 2003;**544**:147.
15. Sloan PA, Palmer RE. Two-electron dissociation of single molecules by atomic manipulation at room temperature. *Nature* 2005;**434**:367.
16. Sakulsermsuk S, Sloan PS, Palmer RE. A new mechanism of atomic manipulation: Bond-selective molecular dissociation via thermally activated electron attachment. *ACS Nano* 2010;**4**:7344.
17. Guo H, Ji W, Polanyi JC, Yang J(SY). Molecular dynamics of localized reaction, experiment and theory: Methyl bromide on Si(111)-7 × 7. *ACS Nano* 2008;**2**:699.
18. Dobrin S, Lu X, Naumkin FY, Polanyi JC, Yang J(SY). Imprinting Br-atoms at Si(111) from a SAM of $CH_3Br(ad)$, with pattern retention. *Surf Sci* 2004;**573**:L363.
19. Harikumar KR, McNab IR, Polanyi JC, Zabet-Khosousi A, Hofer WA. Imprinting self-assembled patterns of lines at a semiconductor surface, using heat, light, or electrons. *Proc Natl Acad Sci USA* 2011;**108**:950.
20. Matta CF, Polanyi JC. Chemistry on a peg-board: the effect of adatom–adatom separation on the reactivity of dihalobenzenes at Si(111)7 × 7 surfaces. *Philos Trans R Soc Lond A* 2004;**362**:1185.
21. Polanyi JC, Rogers D. *Method of molecular-scale pattern imprinting at surfaces.* US Patent specification 2000;**6**,156,393.
22. Polanyi JC, Rogers D. *Method of molecular-scale pattern imprinting at surfaces.* US Patent specification 2001;**6**,319,566.
23. Polanyi JC, Rogers D. *Method of molecular-scale pattern imprinting at surfaces.* US Patent specification **6**,878,417.
24. Polanyi JC, Harikumar KR, McNab IR, Hofer WA. *Method of linear patterning of surfaces.* US provisional patent awarded.
25. Lyubinetsky I, Dohnalek Z, Choyke WJ, Yates Jr. JT. Cl_2 dissociation on Si(100)-(2 × 1): A statistical study by scanning tunneling microscopy. *Phys Rev B* 1998;**58**:7950.

26. Dürr M, Höfer U. Dissociative adsorption of molecular hydrogen on silicon surfaces. *Surf Sci Rep* 2006;**61**:465.
27. Brenig W, Pehlke E. Reaction dynamics of H_2 on Si. *Ab initio* supported model calculations. *Prog Surf Sci* 2008;**83**:263.
28. Dürr M, Hu Z, Biedermann A, Höfer U, Heinz TF. Real-Space Study of the Pathway for Dissociative Adsorption of H_2 on Si(001). *Phys Rev Lett* 2002;**88**:046104.
29. Bratu P, Kompa KL, Höfer U. Optical second-harmonic investigations of H_2 and D_2 adsorption on Si (100) 2×1: the surface temperature dependence of the sticking coefficient. *Chem Phys Lett* 1996;**251**:1.
30. Okamoto Y. Hybrid Density-Functional Calculations of the Reactions of a H_2 Molecule on C, Si, and Ge(001) Surfaces. *J Phys Chem B* 2002;**106**:570.
31. Filippi C, Healy SB, Kratzer P, Pehlke E, Scheffler M. Quantum Monte Carlo Calculations of H_2 Dissociation on Si(001). *Phys Rev Lett* 2002;**89**:166102.
32. Kanai Y, Tilocca A, Selloni A, Car R. First-principles string molecular dynamics: An efficient approach for finding chemical reaction pathways. *J Chem Phys* 2004;**121**:3359.
33. Chabal YJ, Raghavachari K, Zhang X, Garfunkel E. Silanone (Si=O) on Si(100): intermediate for initial silicon oxidation. *Phys Rev B* 2002;**66**:161315(R).
34. Hemeryck A, Mayne AJ, Richard N, Esteve A, Chabal YJ, Djafari Rouhani M, et al. Difficulty for oxygen to incorporate into the silicon network during initial O_2 oxidation of Si (100)-(2×1). *J Chem Phys* 2007;**126**:114707.
35. Perrine KA, Teplyakov AV. Reactivity of selectively terminated single crystal silicon surfaces. *Chem Soc Rev* 2010;**39**:3256–74.
36. Hossain MZ, Yamashita Y, Mukai K, Yoshinobu J. Microscopic observation of precursor-mediated adsorption process of NH_3 on Si(100)c.(4×2) using STM. *Phys Rev B* 2003;**68**:23522.
37. Chung ON, Kim H, Chung S, Koo J-Y. Multiple adsorption configurations of NH_3 molecules on the Si(001) surface. *Phys Rev B* 2006;**73**:033303.
38. Bowler DR, Owen JHG. Molecular interactions and decomposition pathways of NH_3 on Si(001). *Phys Rev B* 2007;**75**:155310.
39. Widjaja Y, Musgrave CB. Indirect adsorbate–adsorbate interactions mediated through the surface electronic structure of the Si(100) surface. *J Chem Phys* 2004;**120**:1555.
40. Wang Y, Hwang GS. Origin of nonlocal interactions in adsorption of polar molecules on Si(001)-2×1. *J Chem Phys* 2005;**122**:164706.
41. Queeney KT, Chabal YJ, Raghavachari K. Role of interdimer interactions in NH_3 dissociation on Si(100)-(2×1). *Phys Rev Lett* 2001;**86**:1046.
42. Hamers RJ, Kohler UK. Determination of the local electronic structure of atomic-sized defects on Si(001) by tunneling spectroscopy. *J Vac Sci Technol A* 1989;**7**:2854.
43. Hossain MZ, Yamashita Y, Mukai K, Yoshinobu J. Microscopic observation of precursor-mediated adsorption process of NH_3 on Si(100)-c(4×2) using STM. *Phys Rev B* 2003;**67**:153307.
44. Okano S, Oshiyama S. A new alternative model of type-C defects on Si(100) surfaces. *Surf Sci* 2004;**554**:272.
45. Zhang Z, Kulakov MA, Bullemer B. New features of C-type defects on the Si(100) surface observed by scanning tunnelling microscopy. *Surf Sci* 1996;**369**:L131.
46. Hata K, Ozawa S, Shigekawa H. Metastable and excited states of the C defects of Si(001). *Surf Sci* 1999;**441**:140.
47. Sobotík P, Ošt'ádal I. Transformations of C-type defects on Si(100)-2×1 surface at room temperature – STM/STS study. *Surf Sci* 2008;**602**:2835. See also: Reply to Comment on

"Transformation of C-type defects on Si(100)-2 × 1 surface at room temperature – STM/STS study" *Surf Sci* 2010;**604**:237.
48. Lim TB, McNab IR, Polanyi JC, Guo H, Ji W. 2010 [in preparation].
49. Guo SY, Ibrahimi M, McNab IR, Polanyi JC. "Early" and "Late" Barriers in Dissociative Attachment: Steering Surface Reaction. *J Phys Chem Lett* 2010;**1**:2600.
50. Dobrin S, Harikumar KR, Polanyi JC. An STM study of the localized atomic reaction of 1,2- and 1,4-dibromobenzene at Si(111)-7×7. *Surf Sci* 2004;**561**:11.
51. Brune H, Wintterlin J, Behn RJ, Ertl G. Surface migration of "hot" adatoms in the course of dissociative chemisorption of oxygen on Al(111). *Phys Rev Lett* 1992;**68**:624.
52. Brune H, Wintterlin J, Trost J, Ertl G, Wiechers J, Behm RJ. Interaction of oxygen with Al(111) studied by scanning tunneling microscopy. *J Chem Phys* 1993;**99**:2128.
53. Komrowski AJ, Sexton JZ, Kummel AC, Binetti M, Weisse O, Hasselbrink E. Oxygen Abstraction from Dioxygen on the Al(111) Surface. *Phys Rev Lett* 2001;**87**:246103.
54. Wintterlin J, Schuster R, Ertl G. Existence of a "Hot" Atom Mechanism for the Dissociation of O_2 on Pt(111). *Phys Rev Lett* 1996;**77**:123.
55. Ellinger C, Stierle A, Robinson IK, Nefedov A, Dosch H. Atmospheric pressure oxidation of Pt(111). *J Phys Conden Matter* 2008;**20**:184013.
56. Herzberg G. Molecular spectra and molecular structure: volume 1, spectra of diatomic molecules. Reprint edition, Krieger, Florida (1989) and references therein.
57. Stipe BM, Rzaei MA, Ho W, Gao S, Persson M, Lundquist BI. Single-Molecule Dissociation by Tunneling Electrons. *Phys Rev Lett* 1997;**78**:4410.
58. McCarty GS, Weiss PS. Footprints of a Surface Chemical Reaction: Dissociative Chemisorption of p-Diiodobenzene on Cu{111}. *J Phys Chem B* 2002;**106**:8005.
59. Lauhon LJ, Ho W. Single-molecule chemistry and vibrational spectroscopy: Pyridine and Benzene on Cu(001). *J Phys Chem A* 2000;**104**:2463.
60. Komeda T, Kim Y, Fujita Y, Sainoo Y, Kawai M. Local chemical reaction of benzene on Cu(110) via STM-induced excitation. *J Chem Phys* 2004;**120**:5347.
61. Dougherty DB, Lee J, Yates Jr. JT. Assembly of linear clusters of iodobenzene dimers on Cu(110). *J Phys Chem B* 2006;**110**:20077.
62. Hofer WA, Leung L, Polanyi JC [to be published].
63. Yang XM, Xu Y, Lee K, Xiao S, Weller D, Kuo D. Advanced lithography for bit patterned media. *IEEE T Magn* 2009;**45**:833.
64. Huo F, Zheng G, Liao X, Giam LR, Chai J, Chen X, et al. Beam pen lithography. *Nat Nano* 2010;**5**:637.
65. Gadzuk JW. Resonance-assisted, hot-electron-induced desorption. *Surf Sci* 1995;**342**:345.
66. Ramsier RD, Yates Jr. JT. Electron-stimulated desorption: Principles and applications. *Surf Sci Rep* 1991;**12**:246.
67. Dobrin S, Harikumar KR, Lim TB, Leung L, McNab IR, Polanyi JC, Sloan PA, Waqar Z, Yang J(SY), Ayissi S, Hofer WA. *Nanotechnology* 2007;**18**:044012.
68. Brown DE, Moffat DJ, Wolkow RA. Isolation of an intrinsic precursor to molecular chemisorption. *Science* 1998;**279**:542.
69. Glueckstein JC, Evans MMR, Nogami J. Growth of Sn on Si(001) at room temperature. *Surf Sci* 1998;**415**:80.
70. Itoh H, Tanabe H, Winau D, Schmid AK, Ichinokawa T. Growth mode and surface structures of the Pb/Si(001) system observed by scanning tunneling microscopy. *J Vac Sci Technol B* 1994;**12**:2086.
71. Tomatsu K, Nakatsuji K, Iimori T, Komori F. STM observation of initial growth of Sn atoms on Ge(001) surface. *Surf Sci* 2007;**601**:1736.

72. Hutchison P, Evans MMR, Nogami J. Initial stages of Mg growth on the Si(001) surface studied by STM. *Surf Sci* 1998;**411**:99.
73. Bakhtizin RZ, Kishimoto J, Hashizume T, Sakurai T. STM study of Sr adsorption on Si(100) surface. *Appl Surf Sci* 1996;**94**:478.
74. Bird CF, Bowler DRA. A spin-polarised first principles study of short dangling bond wires on Si(001). *Surf Sci* 2003;**531**:L351.
75. Shen T-C, Wang C, Abeln GC, Tucker JR, Lyding JW, Avouris Ph, Walkup RE. Atomic-scale desorption through electronic and vibrational excitation mechanisms. *Science* 1995;**268**:1590.
76. Naitoh M, Shimaya H, Nishigaki S, Oishi N, Shoji F. Scanning tunneling microscopy observation of bismuth growth on Si(100) surfaces. *Surf Sci* 1997;**377–379**:899.
77. Naitoh M, Shimaya H, Nishigaki S, Oishi N, Shoji F. Scanning tunneling microscopy observation of Bi-induced surface structures on the Si(100) surface. *Surf Sci* 2001;**482–485**:1440.
78. Owen JHG, Miki K, Bowler DR. Self-assembled nanowires on semiconductor surfaces. *J Mater Sci* 2006;**41**:4568.
79. MacLeod JM, McLean AB. Single 2×1 domain orientation on Si(001) surfaces using aperiodic Bi line arrays. *Phys Rev B* 2004;**70**:41306.
80. MacLeod JM, Miwa RH, Srivastava GP, McLean AB. The electronic origin of contrast reversal in bias-dependent STM images of nanolines. *Surf Sci* 2005;**576**:116.
81. Preinesberger C, Vandre S, Kalka T, Dahne-Preitsch M. The magnetization process and magnetoresistance of exchange-spring bilayer systems. *J Phys D* 1998;**31**:L43.
82. Chen Y, Ohlberg DAA, Williams RS. Nanowires of four epitaxial hexagonal silicides grown on Si(001). *J Appl Phys* 2002;**91**:3213.
83. Nogami J, Liu BZ, Katkov MV, Ohbuchi C, Birge NO. Self-assembled rare-earth silicide nanowires on Si(001). *Phys Rev B* 2001;**63**:233305.
84. Liu BZ, Nogami J. Growth of parallel rare-earth silicide nanowire arrays on vicinal Si(001). *Nanotechnology* 2003;**14**:873.
85. Liu BZ, Nogami J. An STM study of the Si(001)(24)-Dy surface. *Surf Sci* 2001;**488**:399.
86. Ohbuchi C, Nogami J. Holmium growth on Si(001): Surface reconstructions and nanowire formation. *Phys Rev B* 2002;**66**:165323.
87. Katkov MV, Nogami J. Yb and Nd growth on Si(001). *Surf Sci* 2003;**524**:129.
88. Ohbuchi C, Nogami J. Samarium-induced surface reconstructions of Si(001). *Surf Sci* 2005;**579**:157.
89. Harrison BC, Boland JJ. Real-time STM study of inter-nanowire reactions: GdSi2 nanowires on Si(100). *Surf Sci* 2005;**594**:93.
90. Evans MMR, Nogami J. Indium and gallium on Si(001): A closer look at the parallel dimer structure. *Phys Rev B* 1999;**59**:7644.
91. Nogami J, Baski AA, Quate CF. Aluminum on the Si(100) surface: Growth of the first monolayer. *Phys Rev B* 1991;**44**:1415.
92. Lopinski GP, Wayner DDM, Wolkow RA. Self-directed growth of molecular nanostructures on silicon. *Nature* 2000;**406**:48.
93. DiLabio GA, Piva PG, Kruse P, Wolkow RA. Dispersion interactions enable the self-directed growth of linear alkane nanostructures covalently bound to silicon. *J Am Chem Soc* 2004;**126**:16048.
94. Kruse P, Johnson ER, DiLabio GA, Wolkow RA. Patterning of vinylferrocene on H–Si(100) via self-directed growth of molecular lines and STM-induced decomposition. *Nanoletters* 2002;**2**:807.

95. Hossain MZ, Kato HS, Kawai M. Controlled fabrication of 1D molecular lines across the dimer rows on the Si(100)–(2 × 1)–H surface through the radical chain reaction. *J Am Soc Chem* 2005;**127**:15030.
96. Hossain MZ, Kato HS, Kawai M. Fabrication of interconnected 1D molecular lines along and across the dimer rows on the Si(100)(2 × 1)–H surface through the radical chain reaction. *J Phys Chem B* 2005;**109**:23129.
97. Harikumar KR, Lim T, McNab IR, Polanyi JC, Zotti L, Ayissi S, Hofer WA, et al. Dipole-directed assembly of lines of 1,5-dichloropentane on silicon substrates by displacement of surface charge. *Nat Nano* 2008;**3**:222.
98. Dobrin S, Harikumar KR, Jones RV, McNab IR, Polanyi JC, Waqar Z, Yang J(SY). Molecular dynamics of haloalkane corral formation and surface halogenation at Si(111)-7 × 7. *J Chem Phys* 2006;**125**:133407.
99. Harikumar KR, Leung L, McNab IR, Polanyi JC, Lin H, Hofer WA. Cooperative molecular dynamics in surface reactions. *Nat Chem* 2009;**1**:716.
100. Hass KC, Schneider WF, Curioni A, Andreoni W. The chemistry of water on alumina surfaces: Reaction dynamics from first principles. *Science* 1998;**282**:265.
101. Qin F, Magtoto NP, Kelber JA. Moisture-induced instability at the $Al_2O_3/Ni_3Al(110)$ interface: interfacial chemistry. *Mater High Temp* 2004;**21**:193.
102. Lee HS, An K-S, Kim Y, Choi CH. Surface SN_2 reaction by H_2O on Chlorinated Si(100)-2 × 1 surface. *J Phys Chem B* 2005;**109**:10909.
103. Jarvis EA, Chaka AM. Oxidation mechanism and ferryl domain formation on the α-Fe_2O_3 (0001) surface. *Surf Sci* 2007;**601**:1909.
104. http://www.nature.com/nchem/journal/v1/n9/extref/nchem.440-s2.mov.
105. Lim TB, Polanyi JC, Guo H, Ji W. Surface-mediated chain reaction through dissociative attachment. *Nat Chem* 2011;**3**:85.
106. Hofer WA, Fisher AJ, Lopinski GP, Wolkow RA. Electronic structure and STM images of self-assembled styrene lines on a Si(100) surface. *Chem Phys Lett* 2002;**365**:129.
107. Piva PG, DiLabio GA, Pitters JL, Zikovsky J, Rezeq M, Dogel S, et al. Field regulation of single-molecule conductivity by a charged surface atom. *Nature* 2005;**435**:658.
108. Hossain MdZ, Kato HS, Kawai M. Controlled fabrication of 1D molecular lines across the dimer rows on the Si(100)–(2 × 1)–H surface through the radical chain reaction. *J Am Chem Soc* 2005;**127**:15030.
109. Hossain MdZ, Kato HS, Kawai M. Fabrication of interconnected 1D molecular lines along and across the dimer rows on the Si(100)–(2 × 1)–H surface through the radical chain reaction. *J Phys Chem B* 2005;**109**:23129.
110. Maksymovych P, Sorescu DC, Jordan KD, Yates Jr. JT. Collective reactivity of molecular chains self-assembled on a surface. *Science* 2008;**322**:1664.
111. Leung L, Polanyi JC, Hofer WA [to be published].
112. Briner BG, Doering M, Rust H-P, Bradshaw AM. Microscopic molecular diffusion enhanced by adsorbate interactions. *Science* 1997;**278**:257.
113. Komeda T, Kim Y, Kawai M, Persson BNJ, Ueba H. Lateral hopping of molecules induced by excitation of internal vibration mode. *Science* 2002;**295**:2055.
114. Pascual JI, Lorente N, Song Z, Conrad H, Rust H-P. Selectivity in vibrationally mediated single-molecule chemistry. *Nature* 2003;**423**:525.
115. Bartels L, Wang F, Möller D, Knoesel E, Heinz TF. Real-space observation of molecular motion induced by femtosecond laser pulses. *Science* 2004;**305**:648.
116. Backus EHG, Eichler A, Kleyn AW, Bonn M. Real-time observation of molecular motion on a surface. *Science* 2005;**310**:1790.

117. Stépán K, Güdde J, Höfer U. Time-resolved measurement of surface diffusion induced by femtosecond laser pulses. *Phys Rev Lett* 2005;**94**:236103.
118. Gaudioso J, Lee HJ, Ho W. Vibrational analysis of single molecule chemistry: Ethylene dehydrogenation on Ni(110). *J Am Chem Soc* 1999;**121**:8479.
119. Riedel D, Cranney M, Martin M, Guillory R, Dujardin G, Dubois M, et al. Surface-isomerization dynamics of trans-stilbene molecules adsorbed on Si(100)-2 × 1. *J Am Chem Soc* 2009;**131**:5414.
120. Riedel D, Bocquet M-L, Lesnard H, Lastapis M, Lorente M, Sonnet P, et al. Selective scanning tunnelling microscope electron-induced reactions of single biphenyl molecules on a Si(100) surface. *J Am Chem Soc* 2009;**131**:7344.
121. Polanyi JC. Some concepts in reaction dynamics. *Science* 1987;**236**:680.
122. Polanyi JC, Schreiber JL. The reaction of $F + H_2 \rightarrow HF + H$. *Faraday Discuss Chem Soc* 1977;**62**:267.
123. Harikumar KR, Polanyi JC, Zabet-Khosousi A, Czekala P, Lin H, Hofer WA. Directed long-range molecular migration energized by surface reaction. *Nature Chem* 2011;**3**:400.

Chapter 5

Tunnel-Current Induced STM Atomic Manipulation

Peter A. Sloan
Department of Physics, University of Bath, Bath, United Kingdom

1. INTRODUCTION

Scanning tunnelling microscopy (STM) imaging and manipulation of matter at the atomic scale are still, after close to 30 years, revealing new insights into matter at this fundamental level. Presently, atomic manipulation is used to investigate what it itself can do and how these findings affect our understanding of molecular processes on solid surfaces. Atomic manipulation can be broadly split into three main areas depending on the mode of manipulation: (1) mechanical, by the direct interaction of the tip-apex to push, pull or slide adsorbates across a surface; (2) electric field, by the intense electric field ($\sim 10^9$ Vm^{-1}) interacting with a charged, polarized or polarizable adsorbate; and (3) tunnel current, by the electrons that flow in the tunnel current between sample and tip. Mechanical interactions have been extensively studied for both small and large adsorbates predominantly on metal surface at low temperatures,[1–11] whereas there have only been a few reports of electric field-induced manipulation,[12–20] but by far the most active field of research is tunnel current-induced manipulation.[21–30]

At the single atom and molecular level, quantum physics dictates behaviour, yet adsorbates can, within a certain approximation, be regarded as classical objects, especially when mechanically pushing and pulling them. (However, the multi-atom objects created using mechanical manipulation can manifest beautiful quantum behaviour, for example, in quantum corrals.[4,31,32].) On the other hand, tunnel current-induced manipulation is fundamentally quantum mechanical in nature, either through electronic excitation such as negative (or positive) ion resonances or through direct excitation of vibrations, rotations or other quantized degrees of freedom. In this chapter, we will introduce and review the elementary molecular processes that can be induced by the tunnel current from the tip of an STM. The layout of this chapter is as follows. A short experimental section will introduce the STM, the two main types of experimental methods used in atomic manipulation

Frontiers of Nanoscience, Vol. 2. DOI: 10.1016/B978-0-08-096355-6.00005-2

(scanning and single-point) and their associated data analysis schemes. The subsequent three sections each review one specific aspect of STM tunnel current-induced atomic manipulation, desorption of adsorbate atoms and molecules, dissociation of bonds within a molecule and nonlocal atomic manipulation.

2. EXPERIMENTAL METHODS

The timescale of manipulation events, for example, bond breaking, is on the order of a femtosecond, whereas the time resolution of a typical STM is typically microseconds (or worse).[33] There are schemes for combining the time resolution of pulsed laser pump/probe systems with the spatial resolution on an STM, but these are complicated set-ups with no clear interpretation of their results. Therefore, instead of directly 'imaging' the manipulation process, STM images taken before and after a manipulation event give spatial information and the probability per electron of causing an event. The main control parameters of atomic manipulation are the energy of excitation, rate of excitation and precise location of the injection site, that is, the bias voltage, tunnelling current and tip position. Ideally, one would map out this multidimensional parameter space. However, each event is usually a laborious exercise requiring many experimental variables (tip condition, sample cleanliness and vacuum pressure) to be optimized. Therefore, in practice, and as we shall show, often only one or two of these controlling parameters are mapped out. The main experimental methods used to capture quantitative data (i.e. probability per electron as a function of X) are (1) frame-by-frame scanning and (2) single-point current injection.

2.1. Frame-by-Frame Scanning

An area of surface containing the target adsorbates is passively imaged before and after an intervening scan taken with tunnel parameters that promote atomic manipulation. The number of electrons injected into a single adsorbate can be estimated as $n_e = AIt/eS$, where A is the area of one molecule, S the area of the image, I the tunnel current, t the total time of the image and e the electron charge. Comparing the populations of unperturbed adsorbates before (N_0) and after (N) the manipulation scan, and assuming a one electron (or hole) process, the probability per electrons is $P = -\ln(N/N_0)/n_e$. This mode of experimentation gives robust statistics due to the usually high-adsorbate count but cannot give information as to any injection site dependence.

2.2. Single-Point Current Injection

The tip is positioned (usually interrupting a scan) at a precise preselected position on a surface, for example, on top of an adsorbate or at a particular site within a molecule. The feedback loop is disabled, and the voltage ramped to the desired manipulation value. The tunnel current time-trace usually shows a step-wise change at the moment of manipulation. By repeating this process multiple times, a distribution of times-to-manipulation is found with, usually, the form of an exponential decay. The rate of change of the probability P of survival is $dP/dt = -kP$, where k is the probability per second of inducing a manipulation event. This gives $P(t) \propto \exp(-kt)$ which can be transformed to give $P(t) \propto \exp(-k_e It/e)$. The probability per electron, k_e, will be a function of the bias voltage, tip position, adsorbate configuration and, if not a one-electron process, also the tunnel current.

3. DESORPTION

One of the most fundamental of all surface processes is desorption of atoms and molecules from a solid surface. Atomic and molecular desorption is vital to surface catalysis,[34] surface photochemistry[35] and many other surface processes. These desorption processes are induced either by heat or by direct excitation by, for example, light[35] or electrons.[21,36,37] The atomically resolved STM makes an ideal tool to examine, as a passive observer, both types of desorption process. Thermally induced desorption can be studied using the frame-by-frame method, comparing coverage before and after some time interval, allowing calculation of energy barriers and pre-exponential factors for Arrhenius desorption processes.[38–41] Because of its inherent spatial resolution, it is possible to go beyond this level of detail and 'see' what occurs at the surface during one desorption event, for example, the pairing of hydrogen atoms on the Si(100) surface at high temperature prior to recombinative desorption.[42] However, it is the ability of the STM to *cause* atomic and molecular desorption that is of interest here.

Figure 5.1 shows the first reported controlled STM-induced desorption and, in this case, also readsorption.[43] In this seminal work by Eigler *et al.*, an STM tip was positioned 0.5 nm laterally distant from a single Xe atoms adsorbate on the Ni(110) surface at 4 K. A small voltage pulse (Figure 5.1B) induced the Xe atoms to jump (desorb) from surface to tip apex resulting in a higher tunnel current, as the tunnel gap is now reduced by the width of the Xe atom. A second voltage pulse (Figure 5.1D) induced the Xe atoms to return (readsorb) to its original surface binding site. Careful measurements of the probability of Xe desorption/readsorption showed that each Xe atom jump required five tunnelling electrons. This was interpreted as desorption induced by multiple electronic transitions (DIMET).[44–47] As well as investigating the desorption process itself, the ability of the STM to induce atomic and molecular desorption with atomic precision allows machining of nanostructures. There are examples of creating nano-lines[48–50] and of other similar nano-patterns.[51] STM-induced desorption has even been used to create, through the generation of a single dangling bond on the Si(100) surface, a gate voltage that controls the conductance of a molecular transistor.[52]

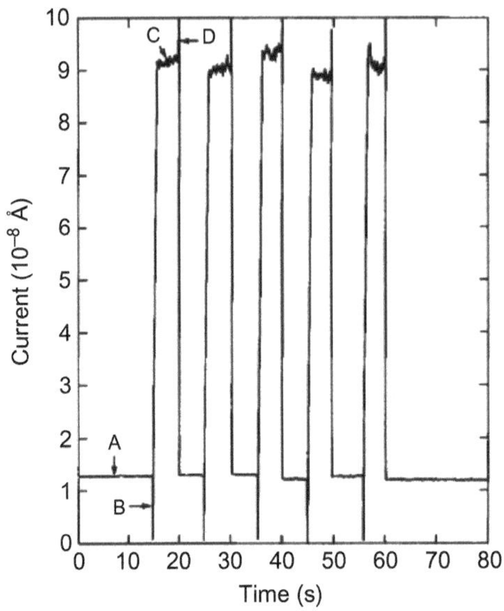

FIGURE 5.1 Time-trace of the tunnelling current as a single Xe atom is induced to shuttle back-and-forth between STM tip (high-current regions C) and Ni(110) surface (low-current region A) at 4 K.[43] B and D indicate the current pulses that induced the atomic desorption and re-adsorption, respectively.

3.1. DIET and DIMET

The two main processes that lead to desorption are desorption induced by electronic transition (DIET) and its complementary partner, DIMET. For a review and exact theoretical details, see Refs. 21 and 53. The most common models for a two-state system are the Menzel–Gomer–Redhead (MGR[54]) and the Antoniewicz[55] models. The key element of these models is to have an excited state (usually an ionic state) that has a different equilibrium bond length to the ground (neutral) state.

Figure 5.2 schematically shows the three potential energy curves associated with three models of DIET, MGR-A (a purely repulsive excited state), MGR-B (a loosely bound excited state) and Antoniewicz (a tightly bound excited state). An excitation to any of these excited state potentials will, while the excitation remains, induce a force to act on the atom or molecule typically for a few femtoseconds,[57] thereby inducing a physical shift of the atom's location. Upon return to the neutral state, the adsorbate will be in a vibrational excited state of the electronic ground state. By this route, vibrational energy (i.e. kinetic energy) can be pumped into the target adsorbate with the possibility of causing desorption. Such nonadiabatic processes are the underlying mechanism for exciting rotations and vibrations which may lead not only to, in this case, desorption but also to diffusion, hopping or bond breaking. The specific outcome depends on the subtleties of the excited state and the neutral state potential surface as well as the coupling of vibrational excitations between the various degrees of freedom of the molecule/surface system.

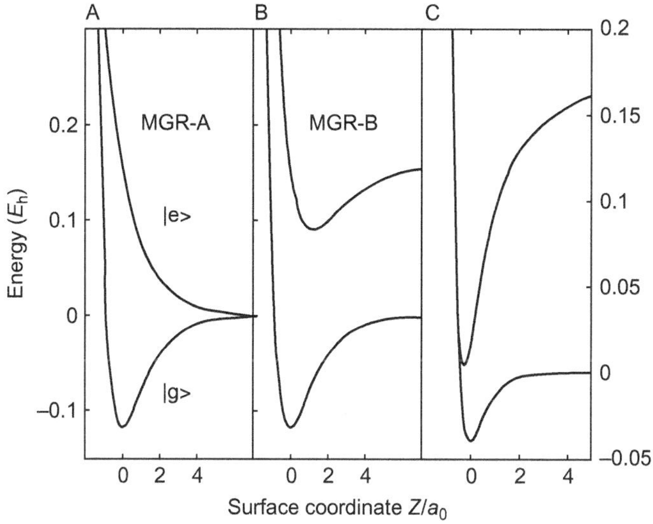

FIGURE 5.2 Models of DIET.[56] (A) Calculated potential energy curves for DIET of hydrogen atoms from the Si(100)-(2 × 1):H surface with purely repulsive excited state due to a σ to σ* transition. (B) As (A) but for a hole resonance excited state. (C) Antoniewicz potential energy curve for the NO on Pt(111) with a negative ion resonance excited state.

The crossover between DIET and DIMET is dependent both on the rates of excitation and relaxation and on the energy of excitation. If the energy of the impinging electron (or hole) from the tip of the STM is lower than the energy barrier to desorption, then, necessarily, more than one excitation event will be required before desorption can occur. If the vibrational relaxation rate is faster of the excitation rate, then each impinging electron will interact with an adsorbate in its ground vibrational state and there is no possibility of multiple excitations combining to overcome an energy barrier. Instead, desorption will be a one electron DIET process. However, if the rate of excitation is higher than the relaxation rate, then, although the DIET process will still be taking place, the dominant desorption process will be 'vibrational ladder climbing' DIMET. In STM manipulation, the number of transitions required for one desorption event can be found by measuring the tunnelling current dependence of the rate of desorption. A DIET process will have a linear dependence, whereas a DIMET process will have a power law dependence on the tunnel current, I^n, where n is the number of transitions (i.e. electrons attachment/detachment events) required to induce desorption. The rate of switching of Eigler's Xe atoms switch has $n \approx 5$, indicating a five-electron process. Persson and colleagues[45] proposed, through computational simulation, that the tunnelling current was resonant with the tail end of the lowest unoccupied molecular orbital of the Xe atom and that there were five vibrational states of the neutral Xe-surface bond agreeing with the DIMET 'vibrational ladder climbing' proposed by Eigler *et al.*

3.2. Si(100)-2 × 1:H

Similar to Eigler's Xe switch, an incoherent multiple excitation mechanism was originally proposed for the much studied system of STM-induced hydrogen desorption from the hydrogen terminated Si(100) surface. The interest in hydrogen-terminated silicon stems from its extensive use in the semiconductor industry to chemically and electrically passivated surface dangling bonds and because such a simple system represents an ideal prototype adsorbate system for the study of the atomic manipulation process itself. In the tunnel regime for both positive sample bias[58–60] and negative,[61] multiple (up to 15 electrons) excitation processes were, it turns out probably incorrectly, identified. At higher bias voltages in the field emission regime, a direct excitation of an electronic σ to σ* transitions within the Si—H system is thought to drive the desorption mechanism.[62,63] Figure 5.3 shows two passive STM images (−1.3 V and 100 pA) taken before (A) and after (B) the three lines indicated had been traced using parameters (+8 V and 10 pA) that induced hydrogen desorption.[58] The bright spots are dangling bond sites where hydrogen atoms have been desorbed from the Si(100)-2 × 1:H surface.

These studies were among the first quantitative STM experiments carried out and have been the subject of some controversy[64,65] as to the validity of the analysis of the experimental data. The crux of the matter was that the range of tunnelling currents used in these pioneering experiments was insufficient to allow an accurate determination of the desorption power law exponent. An extensive study by Soukiassian *et al.*,[49] which examined a wide range of tunnelling currents and tips, found that the number of electrons required to desorb an individual hydrogen atom from Si(100) was in fact ∼1, an order of magnitude smaller than the original reports. Hydrogen desorption from Si (100) is a DIET, not a DIMET, process. This highlights the importance of

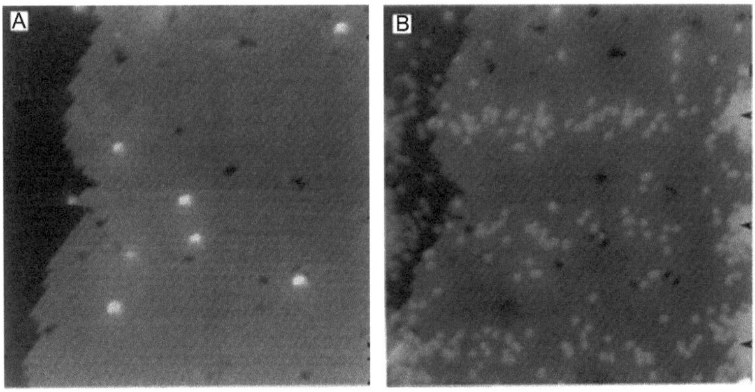

FIGURE 5.3 Hydrogen desorption from Si(100)-2 × 1:H.[58] STM images (200 × 200 Å, −1.3 V, 100 pA) taken before (A) and after (B) three horizontal lines (marked by arrows) were formed by STM-induced desorption at +8 V and 10 pA.

statistics in all atomic manipulation experiments. The Si(100)-2 × 1:H system is still the subject of ongoing research[51,66,67] and has recently been reviewed by Dujardin and co-workers.[68]

3.3. Chlorobenzene/Si(111)-7 × 7

Atomic manipulation has seen a development from atomic adsorbates and simple molecules towards more complex molecules that exhibit a richer array of induced behaviour. However, even relatively simple molecules such as chlorobenzene exhibit a range of possible STM induce reactions, including desorption,[69–71] diffusion[41,72] and C—Cl bond breaking.[73–75] In this section, we review the adsorption and desorption of chlorobenzene on the Si(111)-7 × 7 surface performed by Sloan *et al.* at the University of Birmingham. This prototype system will also be reviewed in the bond dissociation and nonlocal manipulation sections.

When chlorobenzene chemisorbs on the Si(111)-7 × 7 surface, the aromatic nature of the ring is lost and a cyclohexadiene-like 2,5 di-σ bonded butterfly structure is formed.[76] Two sp^3 carbon atoms on opposite sides of the ring bond to an adatom/restatom pair of silicon atoms, leaving two pairs of sp^2 carbon atoms on each wing of the adsorbed molecule. At +1 V sample bias, the signature of these chemisorbed chlorobenzene molecules on Si (111)-7 × 7 is a missing-adatom-like feature. Figure 5.4 shows two STM images, both obtained at +1 V, before (A) and after (B) exposure to chlorobenzene molecules. The darkening of particular adatoms is due to the saturation of the dangling bond upon chemisorption to the adsorbate, a signature observed for many species on the Si(111)-7 × 7 surface.[20,77–80] Imaging at higher bias voltages, to determine the precise orientation of the benzene ring, will be discussed in a later section. Figure 5.5 shows a pair of STM images which were obtained with tunnelling parameters specifically chosen so as not to disturb the system. These images show the same area in each case

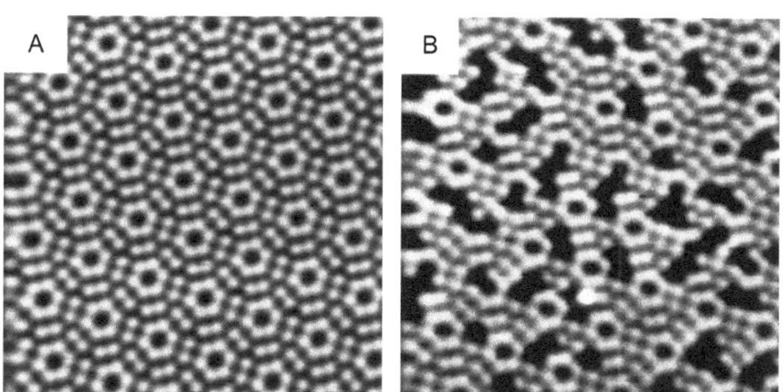

FIGURE 5.4 STM images (150 × 150 Å, +1 V, 50 pA) taken (A) before and (B) after chlorobenzene gas has been dosed onto the surface (150×10^{-8} torr s).

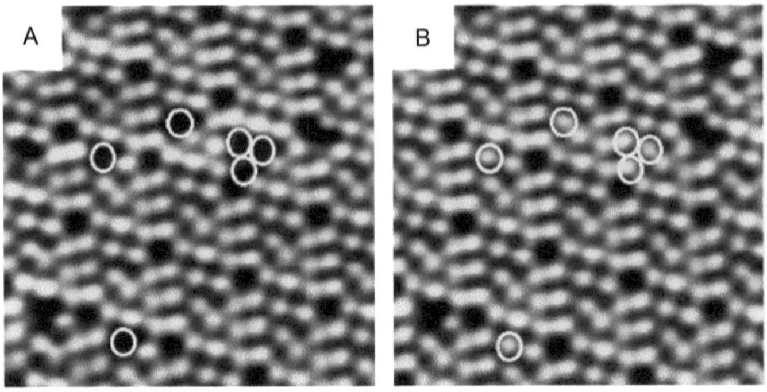

FIGURE 5.5 Desorption of chlorobenzene from Si(111)-7 × 7.[69] STM images (100 × 100 Å, +1 V, 50 pA) taken before (A) and after (B) a manipulation scan (+2.2 V, 50 pA). Circles indicate molecules that were induced to desorb.

and were taken before (A) and after (B) scanning with tunnelling conditions that lead to some desorption of the chlorobenzene molecules (in this case, the manipulation parameters were a sample bias voltage of +2.2 V, a tunnelling current of 50 pA and a tip speed of 280 nm s^{-1}). It is evident that the number of chemisorbed chlorobenzene molecules is reduced in (B) compared with (A), as illustrated by some of the sites circled in the two images.

Measurement of the STM-induced desorption rate as a function of tunnelling current yields a result that is both simple and telling. This can be seen in Figure 5.6 where sample bias voltages of +3 and −2 V were used. In particular, the desorption dependence on current is linear for both voltage polarities and yields a slope in the log–log plot of 0.88 ± 0.09 (+3 V) and 0.89 ± 0.04 (−2 V). The gradient values of the linear fits immediately rule out an incoherent DIMET 'vibrational heating' mechanism, indicating instead that the desorption process is one-electron (or hole) DIET at these bias values. The desorption energy of chlorobenzene from the Si(111)-(7 × 7) is 1.1 eV[76]; therefore, a single tunnelling electron at the biases used has ample energy to induce desorption. The corollary to the linear relationship between desorption rate and tunnelling current is that the probability per electron (often referred to as the yield) will be independent of the tunnelling current. Further, the change in the tunnel current is associated with a change in the tip height (1.74 ± 0.05 Å for +3 V and 2.68 ± 0.07 Å for −2 V); therefore, the probability per electron is also, within experimental error, independent of the tip height. At the higher currents, there is a very slight drop in the desorption yield which may be due to an increased probability of a desorbed molecule 'bouncing' off the tip apex and readsorbing at its original binding site—this would be a desorption event we cannot measure using the frame-by-frame method.

It is worth reiterating the value of this tunnelling current versus rate of manipulation result. If the mechanical presence of the tip did play an active

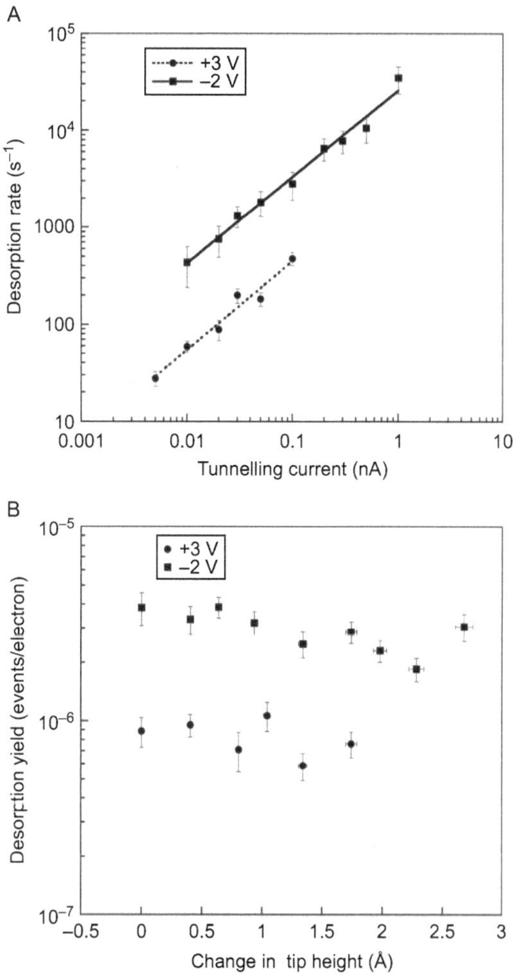

FIGURE 5.6 Rate of chlorobenzene desorption from Si (111)-7 × 7.[69] (A) Tunnel current dependence of desorption at +3 and −2 V with power law fits. (B) Desorption yield (probability per electron) as a function of the change in the tip height. In both cases, the −2-V data have been shift up by an order of magnitude to ease viewing.

role in the desorption, one would expect that the interaction would change over the range of tip heights probed and, further, that this would be manifest as a substantial change in the desorption yield. A similar argument can be made about the electric field affecting the desorption process. Assuming an absolute tip height from the surface in the region of 6 Å,[19,81,82] then the tip-height changes associated with the change in tunnel current (at fixed voltage) are substantial proportions of the total tip height. This should consequently lead to substantial changes in the electric field as the tip-height changes. Any process that had a dependence on the electric field should therefore be affected by this change in the tip height, but no significant change of the yield was observed over the range of tip heights probed. We can therefore rule out processes in which the electric field enhances desorption by, for example, reducing the barrier to desorption and processes in which the electric field shifts in energy the electronic states (i.e. a Stark shift). That is not to say that these effects are not happening to some degree, but that the desorption process is, within error, insensitive to them.

In the system of benzene on Si(100),[38,83–88] which in terms of binding geometry and energy mimics chlorobenzene/Si(111)-7 × 7, Wolkow and co-workers[86] compared their STM-induced desorption results, similar to those found for chlorobenzene/Si(111)-7 × 7, with *ab initio* calculations. Figure 5.7 shows two calculated potential energy surfaces for (A) the neutral state of benzene on Si(100) and (B) the anion state. The abscissa axis described the degree of ring bending and the ordinate axis described a desorption coordinate. Figure 5.7A clearly shows the bonding state of the neutral molecule and the energy barrier to desorption. The excited state potential energy surface has a different location of its bound state (cf. the potential energy curves of Figure 5.2), thus an excitation from neutral to excited state and subsequent neutralization will result in vibrational excitation of the molecule, specifically the ring bending mode, which in turn couples efficiently to the desorption

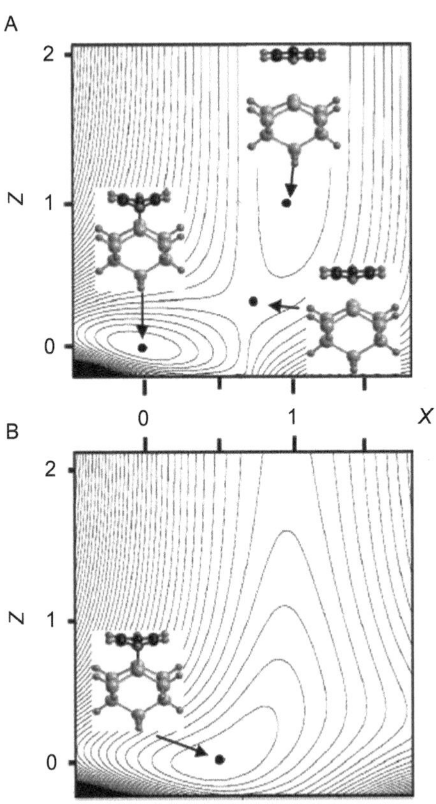

FIGURE 5.7 DIET of benzene on Si (100).[86] Calculated potential energy surfaces for (A) the neutral state and (B) the anion state of the benzene adsorbate. The dimensionless coordinates X refers to a ring-bending mode and Z to the desorption coordinate. The black dots mark the location of stationary states with associated schematic diagrams.

coordinate leading to molecular desorption of a neutral benzene molecule. The complex interaction of an adsorbate with a surface prohibits, to an extent, the *a priori* knowledge of what outcome a particular STM excitation may have. That is not to say that general rules cannot be formed (e.g. unsaturated molecules on silicon surfaces desorb under electron excitation[89]), but that each individual report of atomic manipulation, more often than not, exhibits behaviour specific to that molecule/surface system.

4. INTRAMOLECULAR BOND DISSOCIATION

In 2000, Rieder and co-workers used an STM to instigate and control all the steps required to perform a single chemical reaction.[90] Tunnel electrons dissociated the C—I bond of two iodobenzene molecules; the STM tip was then used to 'mechanically' drag the two benzene radicals together before a second tunnel current pulse initiated fusion between the two rings to form a single biphenyl molecule (see Figure 5.8). There are three clear steps in this example of an Ullmann reaction,[91] bond breaking, mechanical manipulation and bond making. In this section, we will review that first bond-breaking step induced by the STM tunnel current.

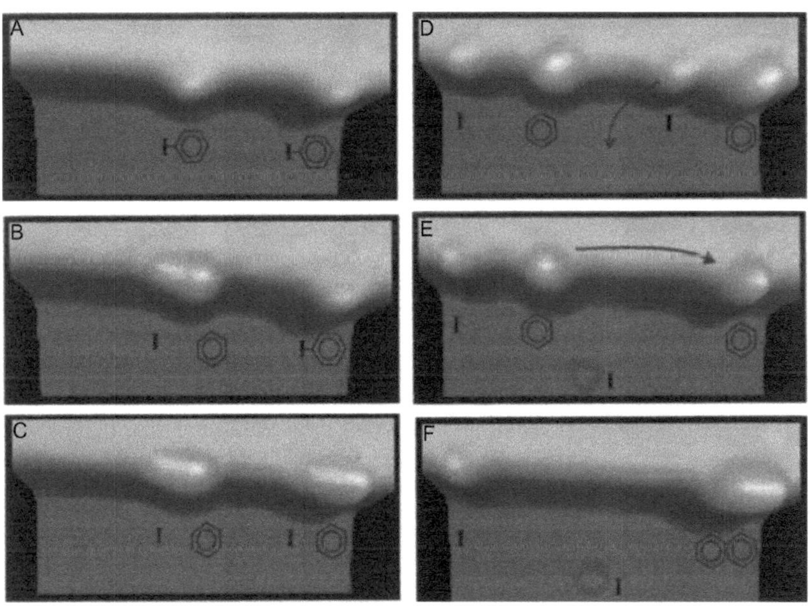

FIGURE 5.8 Inducing all steps in a single Ullmann reaction.[90] STM images (70 × 30 Å, +100 mV, 530 pA) showing (A) two iodobenzene molecules at a step edge on the Cu(111) surface; (B) after STM-induced dissociation of one C—I bond; (C) after dissociation of the other C—I bond; (D) showing the intended removal of one of the iodine atoms by mechanical manipulation; (E) showing the intended mechanical manipulation to bring the two benzene radicals into proximity to each other; and (F) after a tunnel current pulse induces ring fusion.

With the advent of STM, it is possible to break individual bonds within a single preselected molecule. But the STM can do more than that as the imaging capabilities of the STM allows the atomic location and configuration of the molecule before and after dissociation to be known. This spatial information, unique to STM manipulation, gives further information as to the precise bond-breaking mechanism. Here, we review the STM-induced bond dissociation of O_2/Pt(111)[92] performed by Stipe *et al.*[24] We then return to the chlorobenzene/Si(111)-7 × 7 system and the tunnel current-induced dissociation of the C—Cl bond which highlights the power of atomically resolved images before and after dissociation.[75] The final example is from the Polanyi reaction dynamics group of a complex STM-induced dissociation mechanism that involves the cooperative dynamics of two neighbouring molecules.[93]

4.1. O_2/Pt(111)

The first controlled example of STM-induced bond breaking, and still one of the most complete, by Stipe *et al.*[24] was of oxygen on the Pt(111) surface at low (40–150 K) temperature. A very stable STM[94] (drift in the z-direction of 0.001 Å min^{-1}) was used to position the STM tip precisely over an O_2 adsorbate. With the feedback switched off and the tip height changed to a preset value to generate the required current, the voltage was ramped to a preset value and the tunnelling current measured as a function of time (Figure 5.9B). The sharp increase of the current at $t=0$ signals the voltage ramp to the preset value (+0.3 V), and the sudden current drop at 30 ms indicates the precise time the O_2 dissociated into its two constituent oxygen atoms. Figure 5.9A shows two oxygen molecules (indicated by F) one of which was induced to dissociate into the individual oxygen atom products (Figure 5.9C, h and f). Figure 5.9D is the result after the second molecule has undergone bond dissociation. The current and bias dependence of the dissociation rate (Figure 5.10) was measured by examining the distribution of times taken for the dissociation of 152 oxygen molecules. Recall that the rate of a current driven process varies as I^n, with n the number of electrons involved in the process. When the tunnelling electron has more energy than the barrier to dissociation, only one electron is needed and hence the slope for +0.4-V bias voltage is approximately 1 (0.8 ± 0.2). At a reduced bias of +0.3 V, two electrons ($n=1.8\pm0.2$) are required to cause dissociation, and if reduced further to +0.2 V, then three electrons are needed per dissociation event ($n=2.9\pm0.3$).

The proposed mechanism for this dissociation is shown schematically in Figure 5.11A and B. For a bias of 0.4 V, a single-step process is dominant over a multiple excitation process because the vibrational relaxation rate 4×10^{12} s^{-1} is faster than the excitation rate 6×10^{11} s^{-1}.[92] However, at a bias voltage of +0.3 V, a single electron has insufficient energy to dissociate the molecule and two electron scattering events are required. At a lower bias of +0.2 V, three electron scattering events are required to break the bond.

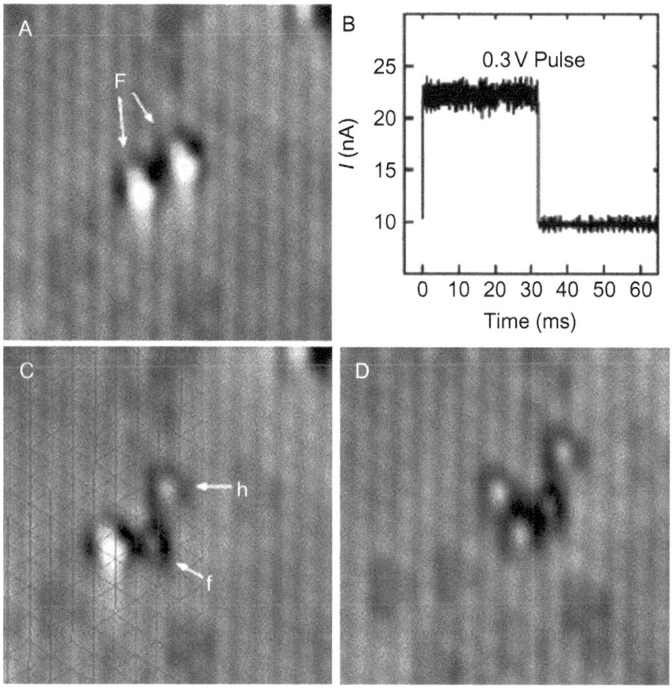

FIGURE 5.9 Dissociation of oxygen on Pt(111).[24] STM images (50 × 50 Å, 25 mV) taken of (A) two oxygen molecules, (C) after current injection as show in the tunnel current time-trace (B) to dissociate a single oxygen molecule into its two oxygen atoms and (D) after subsequent dissociation of the second oxygen molecule.

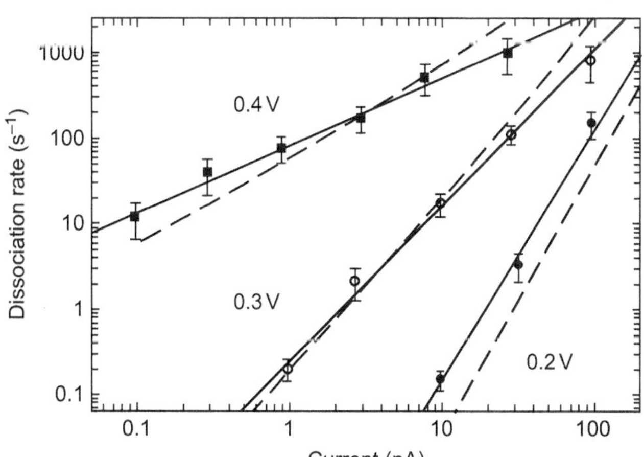

FIGURE 5.10 Dissociation rate of O_2 on Pt(111) as a function of injection tunnel current[24] for three bias voltages. Power law fits give the number of electrons involved in each experiment (see main text for details).

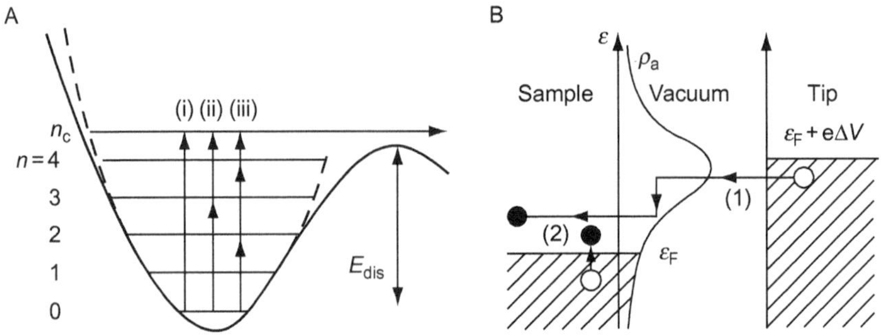

FIGURE 5.11 Mechanism of O_2 dissociation.[24] (A) Schematic potential energy curve with vibrational energy levels showing the vibrational excitation induced by (i) a 0.4-eV electron, (ii) 0.3 eV and (iii) 0.2 eV. (B) Tunnelling electrons may cause vibrational excitation of the O—O bond (1), the substrate may quench this excitation by an electronic excitation (2).

(Presumably between scattering events, some of the vibrational excitation is lost to the surface.) If an even lower voltage was used, then excitation of a rotational mode of the O_2 can occur,[95] spinning the molecule instead of dissecting it.

4.2. Chlorobenzene/Si(111)7 × 7

Figure 5.12 shows a series of sequentially acquired STM images that capture the C—Cl bond dissociation of a single chlorobenzene molecule in the Si (111)-7 × 7 surface. Figure 5.12A and B at +1 and +2 V allows the identification of three chlorobenzene molecules at sites α, γ and δ. During the image in Figure 5.12C taken at +3 V, two molecules at γ and δ were induced to desorb. At +4 V, Figure 5.12D, there is a 'half sunrise' feature. This is the signature of STM-induced C—Cl bond breaking. As the STM scanned from the image bottom to top at exactly the location marked in Figure 5.12I and J, the C—Cl bond breaks and, instead of imaging a chemisorbed chlorobenzene molecule at +4 V, a chemisorbed chlorine atom at site β is found. Figure 5.12 (F–H, +3 to +1 V) allows the identification of the ejected chlorine atom (dark at +1 V, but bright at +2 V and above). The atomic resolution of the STM allows the precise identification of the parent (chlorobenzene molecule) and daughter (chlorine atom) spatial relationship, that is, the distance of daughter from parent and the angle of ejection relative to the original C—Cl bond direction.

Both HREELS experiments[76] and DFT calculations[69] indicate that the carbon atom of the C—Cl bond in the parent chlorobenzene molecule is not one of the two carbons that form covalent bonds to two surface silicon atoms. Figure 5.13 shows STM images where both the bonding silicon adatom and restatom pair can be determined. The signature of the bonding adatom is simply a 'missing' adatom black spot at +1 V, and the signature of the bonding

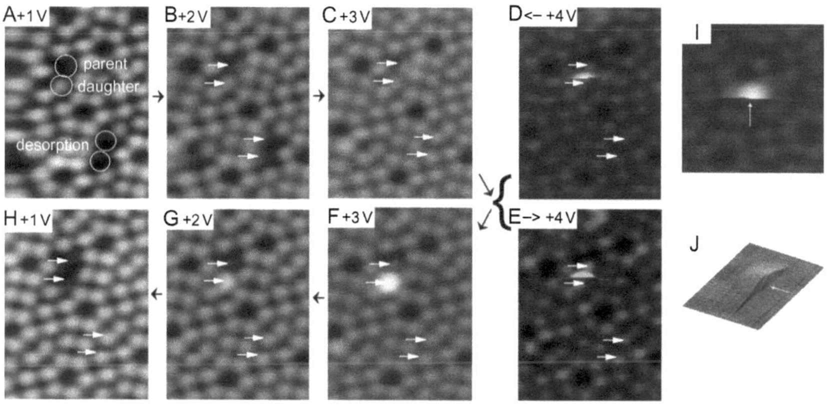

FIGURE 5.12 Dissociation of a single chlorobenzene molecule.[75] Sequence of STM images (100 pA) taken at (A) +1 V, (B) +2 V, (C) +3 V, (D, E) +4 V, (F) +3 V, (G) +2 V and (H) +1 V. Two chlorobenzenes that desorb are marked at sites γ and δ. The chlorobenzene molecule at site α was induced to dissociate producing a chlorine atom at site β. (I, J) Two- and three-dimensional STM images (+4 V) showing a 'half sunrise' signature of C—Cl bond breaking while scanning.

restatom is a slight bright feature located at the site of the bonding restatom at a bias of +2 V. By this means, we can, within a fourfold symmetry (see inset of Figure 5.13G), determine the radial distance and angular distribution (Figure 5.13G and H) of the daughter chlorine atoms from their parent chlorobenzene molecules. When high current (500 pA) is used during the dissociation process, there is a reasonably isotropic spread of daughter angles, whereas at low current (100 pA), the distribution is peaked at 45°. At the higher current, the chlorine atoms were also found further away from the parents, up to ~50 Å, than at the lower current where chlorine atoms are predominantly found at < 10 Å from the parent site. Using high current during the C—Cl dissociation appears to give more kinetic energy to the ejected chlorine atom allowing it to travel further and scatter. Hence they lose their memory of the original C—Cl bonding angle. At the lower current, however, the chlorine atoms appear to drop off and be steered to the nearest available silicon site which lies at 45°, some 15° away from the C—Cl bond angle of 60°. The tunnel current dependence of the rate of C—Cl dissociation was found to be quadratic; two electrons were required to break one C—Cl bond even although each 4 eV incident electron has ample energy to break the 1.9 eV C—Cl bond in the chemisorbed chlorobenzene molecule.[96]

In gas phase dissociative electron attachment of chlorobenzene, an electron first attaches to the π* orbital of the ring and, by subsequent loss of its excess energy to ring vibrations, lowers the symmetry barrier between the π* of the ring and the σ* orbital of the C—Cl bond allowing the electron to populate that C—Cl repulsive state.[97] This generates a negative chlorine

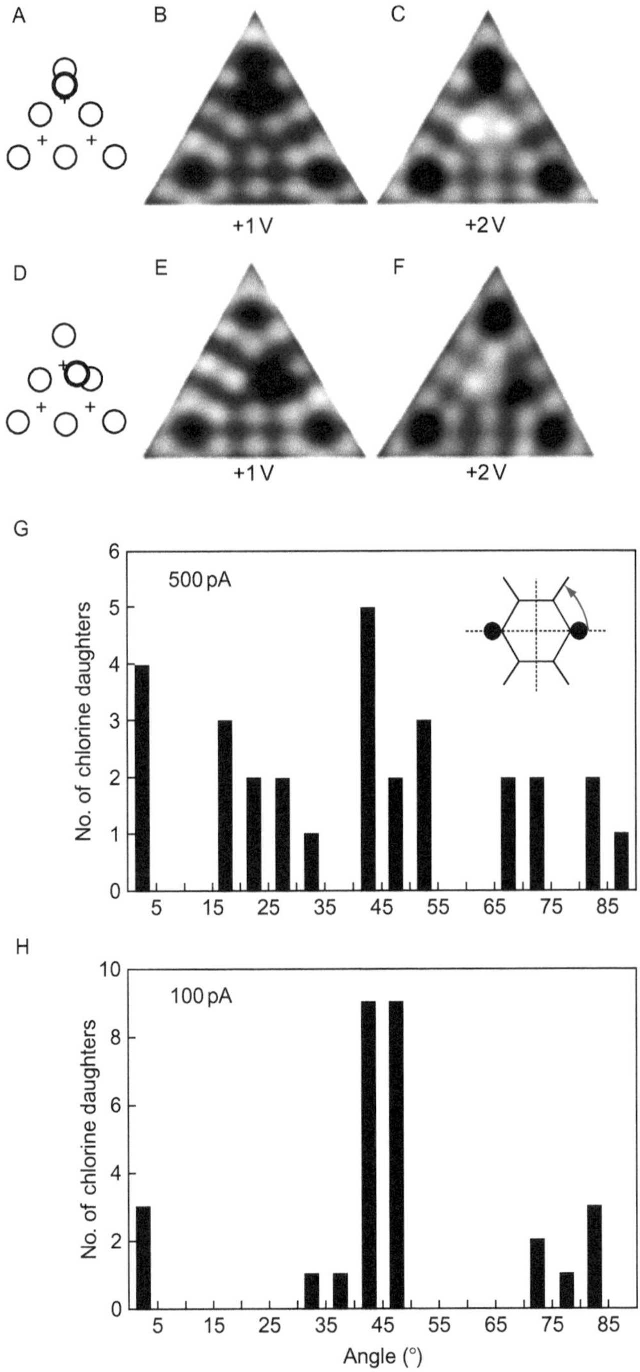

FIGURE 5.13 Angle resolved C—Cl dissociation.[75] (A–F) STM images and schematic diagrams of a chlorobenzene molecule bonded to a corner adatom/restatom pair (A–C) and to a middle adatom/restatom pair (D–E). Angular distribution of ejected chlorine atoms relative to the adatom/restatom axis (see inset g) for a dissociation current of (G) 500 pA and (H) 100 pA.

ion with near zero kinetic energy. A similar one-electron process is unlikely on a surface, as the negative-ion lifetime will be on the order of 1 fs,[85,98,99] insufficient time for one electron to instigate a one-electron gas phase process. Instead, we proposed the following mechanism for STM-induced C—Cl bond breaking: (1) an electron transiently attaches to the π^* of the molecule causing vibrational excitation with some molecules desorbing, (2) the vibration excitation decays, (3) a second electron transiently attaches to the π^* but can, if vibrational excitation still exist within the molecule, transfer to the C—Cl σ^* leading to (4) C—Cl dissociation and ejection of a chlorine atom. If the time interval between the two electrons is sufficiently brief, then the molecule may still be highly vibrationally excited and the second electron can directly populate the C—Cl σ^* and break that bond. The excess energy of the second electron is presumably converted to kinetic energy of the ejected chlorine atom. If, however, the time between impinging electrons is longer, then some of the second electron's excess energy will be channelled into vibrational energy before that same electron populates the C—Cl σ^* and breaks the bond, thus ejecting the chlorine atoms with little or no kinetic energy.

4.3. Fluoropentane/Si(100)

One of the aims of nanoscience is to create pre-designed patterns at the atomic scale.[100,101] One possible route to large-scale patterning is to use a self-assembled monolayer of physisorbed molecules to generate the required pattern, followed by the imprinting of that fragile pattern by induced chemical reaction. This method relies on localized atomic reaction so that there is high fidelity in the imprinted pattern.[73] In the previous example, we saw that chlorine atoms liberated from chlorobenzene molecules can escape by up to 50 Å; here, we review an STM-induced reaction that is, by its cooperative nature, strictly local to the parent site.[93]

Figure 5.14 presents STM images (B–D), schematic diagrams (A and E) and simulated STM images (F–H) of the cooperative reaction of two fluoropentane molecules on the Si(100) surface initiated by charge injection from an STM tip. Fluoropentane molecules were found to self-assemble in pairs with a head-to-head configuration. The C—F bonds of each molecule locate over the atoms of a silicon dimer. This self-assembly pairing was most probably due to a surface-induced dipole interaction.[102–104] Injecting charge at the position of one of the silicon dimer atoms caused reaction, with the final outcome that two fluorine atoms chemisorbed to the two atoms of the silicon dimer. Although this reaction involved the breaking of two C—F bonds, it required only a single electron (or hole). The voltage dependence of reaction had thresholds at +1.4 and −2.4 V. These correspond to the π^* and π state of the silicon dimer pair. Heat was also found to drive this cooperative molecule reaction.

By comparing STM images with meticulous DFT calculation, the following reaction mechanism was proposed. As the fluorine atoms of one of the fluoropentane molecules approach its underlying silicon atom, it overcomes

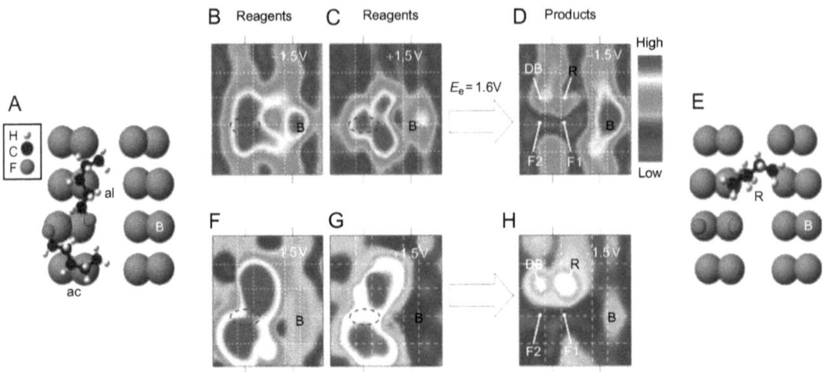

FIGURE 5.14 Cooperative reaction of two fluoropentane molecules on Si(100).[93] (A) and (E) Schematics of molecular configuration before and after reaction. Filled (B) and unfilled (C) STM images before reaction and (D) filled STM image after reaction (19 × 19 Å, 200 pA). Simulated STM images (F–H) equivalent to the STM images.

a 1.4 eV barrier and the organic tail restructures its position and desorbs as the fluorine atom completes its transfer to the silicon atom. At this juncture, the silicon dimer is split creating a dangling bond (i.e. a radical) at the location of the C—F bond of the second molecule which, with unit probability, breaks to generate a second chemisorbed fluorine atom at the other silicon atom of the dimer. Thus, a single electron generates two chemisorbed fluorine atoms at neighbouring silicon atoms sites. Such a radial-mediated reaction is an example of a, albeit limited in this case, chain reaction induced by STM.[105,106] The key is the retention of the original physisorbed position of the two C—F bonds in the final position of the chemisorbed fluorine atoms: local atomic reaction. There are many other examples of STM-induced bond dissociation which rely on the atomic resolution of the STM and excitation by the tunnel current to both view and induce chemical reaction.[107–112]

5. NONLOCAL MANIPULATION

In most of the studies reviewed and discussed so far, the surface is, to a large extent, regarded as a passive substrate or a 'peg board' for entrapping atoms and molecules.[113] There is, however, a cornucopia of surface physics, from surface electronic states to surface reconstruction, that will have a degree of influence on atomic manipulation. To prevent such complications and effectively decouple molecular adsorbates from the surface, a technique of using an atomically thin insulating layer between adsorbate and conducting surface has been developed with striking imaging and manipulation results.[110,114–116] Here, however, we introduce a relatively new mode of atomic manipulation that explicitly exploits the surface properties of particular systems to extend the spatial range of atomic manipulation, namely, nonlocal manipulation.

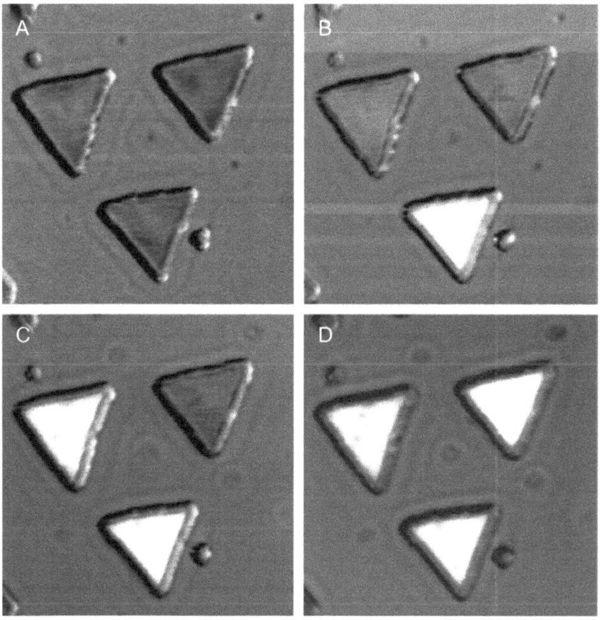

FIGURE 5.15 Nonlocal dehydrogenation of Co islands on Cu(111).[118] STS maps (460 × 460 Å) taken at −0.3 V and 2 nA showing (A) the hydrogenated Co nanoislands. (B) After charge injection and nonlocal hydrogen desorption within the island of the lower island, (C) the island on the left and finally, (D) the island on the right.

Instead of manipulation exclusively occurring in the tunnel junction, manipulation can occur many nanometres distant from the injection site. Nonlocal manipulation has been found for the diffusion of water clusters on Ag (111),[117] the dehydrogenation of Co islands on Cu(111)[118] (see Figure 5.15) and in chemical overlayers on Ag(111) and Au(111).[119]

On the Si(111)-7 × 7 surface, nonlocal manipulation is found in C_{60} monolayers,[120] chemisorbed chlorine atoms[121–125] and for physisorbed chlorobenzene molecules.[72] Nonlocal manipulation has also recently been the subject of a 'perspective' article.[126] Here, we review two nonlocal works, the dissociation of disulphide molecules on Au(111)[127] and the nonlocal desorption of chlorobenzene from the Si(111)-7 × 7 surface.[99] There are also examples of nonlocal manipulation that instead of using the properties of the native surface, take advantage of the local electronic disruption around an adsorbate (typically a few nanometres) to extend the distance over which manipulation occurs.[128,129]

5.1. CH_3SSCH_3/Au(111)

To produce a flow of electrons between tip and surface in an STM, a bias voltage is applied. The close proximity of the electrically biased tip to the surface creates an intense electric field. It is therefore not possible to entirely separate the effect of the tunnel current from the effect of the electric field. Maksymovych *et al.* used nonlocal manipulation to expose the effect of the *E*-field by comparing the dissociation of CH_3SSCH_3 molecules on the Au (111) that occur in the tunnel junction with those remote from that region.[127]

Figure 5.16A and B shows high-resolution STM images and associated schematics of the dissociation of one CH_3SSCH_3 molecule into two SCH_3 fragments. Figure 5.16C presents a large scan of an Au(111) surface decorated with CH_3SSCH_3 after an injection of charge (+2.5 V, 3×10^9 electrons) into the centre of the image. An area of dissociated molecules is clearly visible centred on the injection site. About 1000 molecules were dissociated during each injection experiment including all the molecules within 20 nm of the injection site. This large number of events gives statistical robustness. The authors found that within experimental error, the same number of dissociation

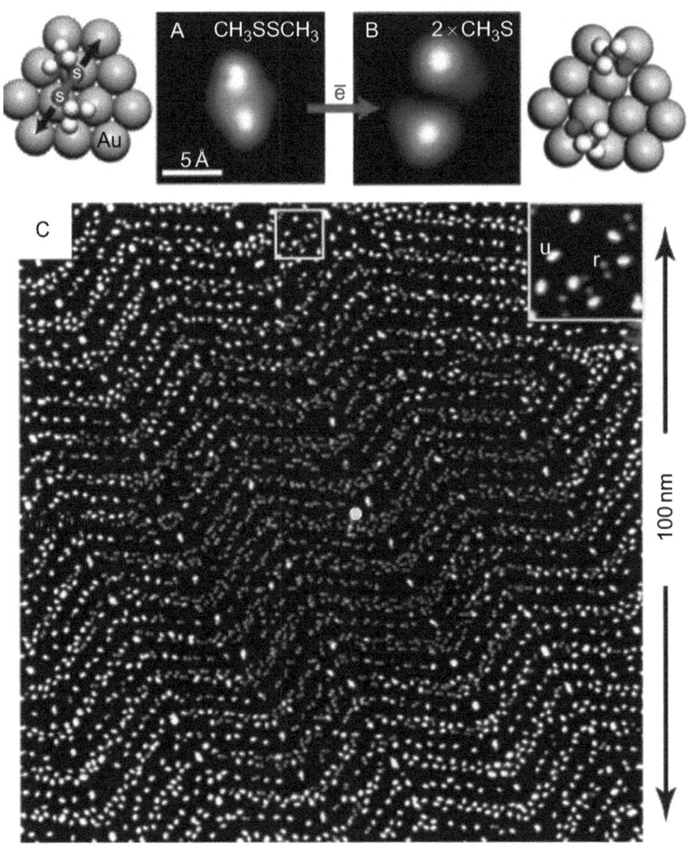

FIGURE 5.16 Nonlocal dissociation of CH_3SSCH_3 on Au(111).[127] STM images and schematics before (A) and after (B) induced S—S bond dissociation. (C) Large area image showing nonlocal dissociation after current injection (+2.5 V, 1 nA, 200 mS) at centre of image.

events occurred for many such injection events. The high threshold voltage of +1.4 V suggested that the dissociation mechanism was electronic and not, as usual for metal substrate systems, due to vibrational or rotational excitations. To reveal the role of the tunnel junction electric field, they compared the final configuration of dissociated fragments in the tunnel junction and those remote from it. Subtle differences in the configuration of the fragments were found suggesting that role of the E-field was not to drastically change the manipulation process but was instead to delicately alter its final outcome. Similar directional *E*-field influence was observed for STM-induced diffusion of CH_3S on Cu(111).[15]

5.2. Chlorobenzene/Si(111)-7 × 7

Semiconducting surfaces have well-defined electronic states both in terms of energy and in terms of spatial distribution. Yet the underlying surface state(s) that mediate nonlocal manipulation on, for example, the Si(111)-7 × 7 surface remain unknown even though there have been several reports of nonlocal manipulation on this surface.[72,120–125] If a clear understanding or identification of the surface state was made, then nonlocal manipulation could be controlled and engineered, for example, by quenching the mediating surface state in certain regions to break the isotropic nature of nonlocal manipulation. This would open the way to (relative!) mass production, but still with atomic spatial resolution, using atomic manipulation. To uncover the critical role of the surface and identify the surface state that links injection site and target molecule, we studied the nonlocal desorption of chlorobenzene from the Si(111)-7 × 7 surface at room temperature.

Figure 5.17 shows a pair of STM images taken before (A) and after (B) injection of electrons at the centre of the 'x'. To inject charge the scan was halted at a predetermined position, the voltage ramped quickly to a preset voltage (+3.6 V) for a preset time (4.46 s) at the scan current (250 pA). The post-injection image (B) shows an area ~150 Å in diameter surrounding the injection site depopulated of chlorobenzene molecules. Nonlocal desorption was found for injection of both electrons and holes. Neither grain boundaries nor steps in the surface halted the nonlocal desorption. Figure 5.18 presents the probability per electron that impinges on a molecule of inducing desorption as a function of radial distance from the injection site (see Ref. 99 for analytical details). The decay is fitted with a single exponential with best fit parameters of $k_e = (4 \pm 0.1) \times 10^{-8}$ and $\lambda = 173 \pm 5$ Å. The two parameters k_e and λ characterize the nonlocal effect, k_e the probability of desorption per electron that impinges on a molecule and λ the decay length of the effect. Both parameters are independent of the tunnelling current and of the duration of the injection.

To confirm that the injected current drives the nonlocal desorption, a family of nonlocal decay curves were taken with the same injection voltage and

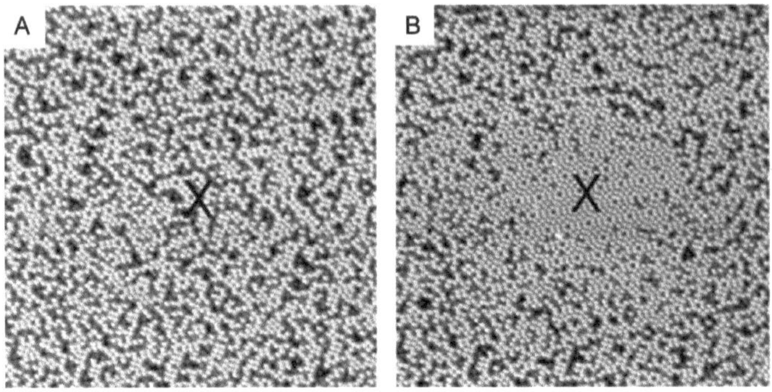

FIGURE 5.17 Nonlocal desorption chlorobenzene from Si(111)-7 × 7.[99] STM images (512 × 512 Å, +1 V, 250 pA) taken before (A) and after (B) charge injection (+3.6 V, 250 pA, 4.46 s) at the centre of the 'X'.

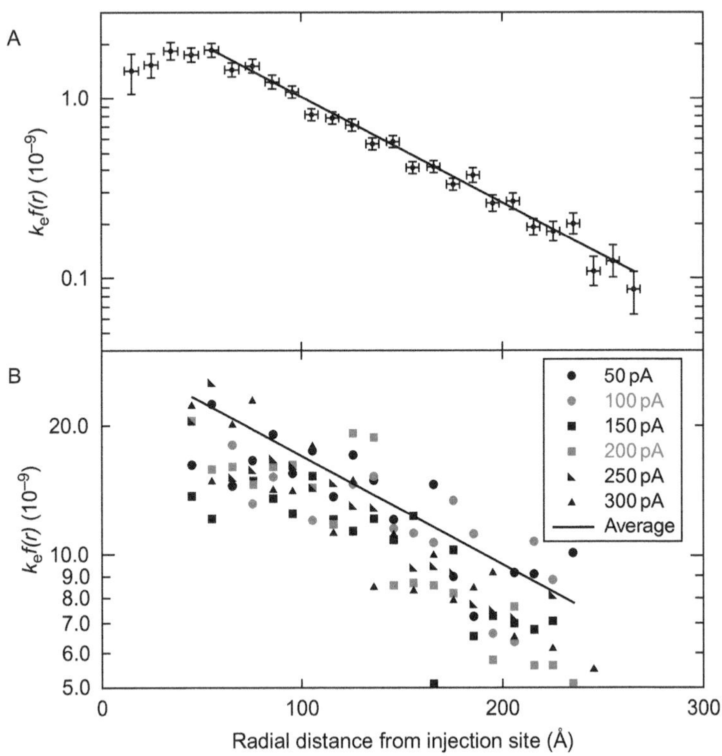

FIGURE 5.18 Quantifying nonlocal desorption chlorobenzene from Si(111)-7 × 7.[99] (A) Nonlocal desorption as a function of radial distance away from injection site with exponential fit. (B) Suite of six nonlocal radial curves taken with different injection currents and times, but each with the same total injection charge and voltage.

total injected charge, but with differing tunnel currents and injection times. These radial decay curves were found to be effectively identical confirming that each nonlocal desorption event is a one-electron process. Following a similar analysis to Ref. 127, the number of electrons per nonlocal desorption event was calculated as 0.90 ± 0.03 (cf. our previous measurement using the scanning method of 0.88 ± 0.09).

To examine the electronic structure of the Si(111)-7×7 surface, we measured 128 equally spaced scanning tunnelling spectra along the high-symmetry line from corner hole (CH) to CH of a unit cell. Figure 5.19B shows the $(dI/dV)/(I/V)$ map generated from averaging 4096 spectra. The known +0.5 V dangling bond (usually labelled U_1) and +1.7 V backbond states (U_2) are evident, but both states are below the +2.1 V threshold found for nonlocal desorption.[130] Instead, we find at + 2.1 V a state that is predominantly located at the CH and restatom sites (faulted restatom, FR and unfaulted restatom, UR). Figure 5.19C presents the probability per impinging electron injected at +2.7 V as a function of injection site across a Si(111)-7×7 unit cell in comparison to the integrated local density of states (LDOS) from +2.1 to +2.7 V. The preference for injection at the CH sites and the faulted half of the unit cell in general are well matched by the integrated LDOS. We therefore identify the surface state at +2.1 V that is predominantly localized on the CH sites and the faulted half of the unit cell as the surface state that transports the injected charge from injection site to distant target molecules.

This type of nonlocal molecular manipulation process has implications for atomic manipulation, electron beam and photochemistry on the Si(111)-7×7 and other surfaces. Threshold voltages may be set by the surface electronic states and not, as is generally assumed, by the molecular adsorbate states. The molecular threshold for desorption can act as a high-pass energy filter, selecting which current propagating surface state couples to molecular manipulation, for example, here the U_1 and U_2 surface states are rejected. The probability of manipulation per electron extracted from STM scanning experiments overestimates the true probability by nearly 2 orders of magnitude. This will impact the resonance state lifetimes required to generate a given manipulation probability in calculations,[85,98] which sometimes seem unfeasibly large. The same consideration applies to processes driven by photoelectrons or secondary electrons.[35]

As the level of experimental and analysis sophistication increases, so too does the range of parameters that atomic manipulation can explore. Recent advances aiming to combine STM with atomic force microscopy, such as the qPlus or Kolibri sensor, may give unprecedented insight into the making and breaking of individual chemical bonds. Even after almost 30 years since Eigler's 'IBM' nanoadvert, atomic manipulation is making often surprising discoveries about the building blocks of matter.

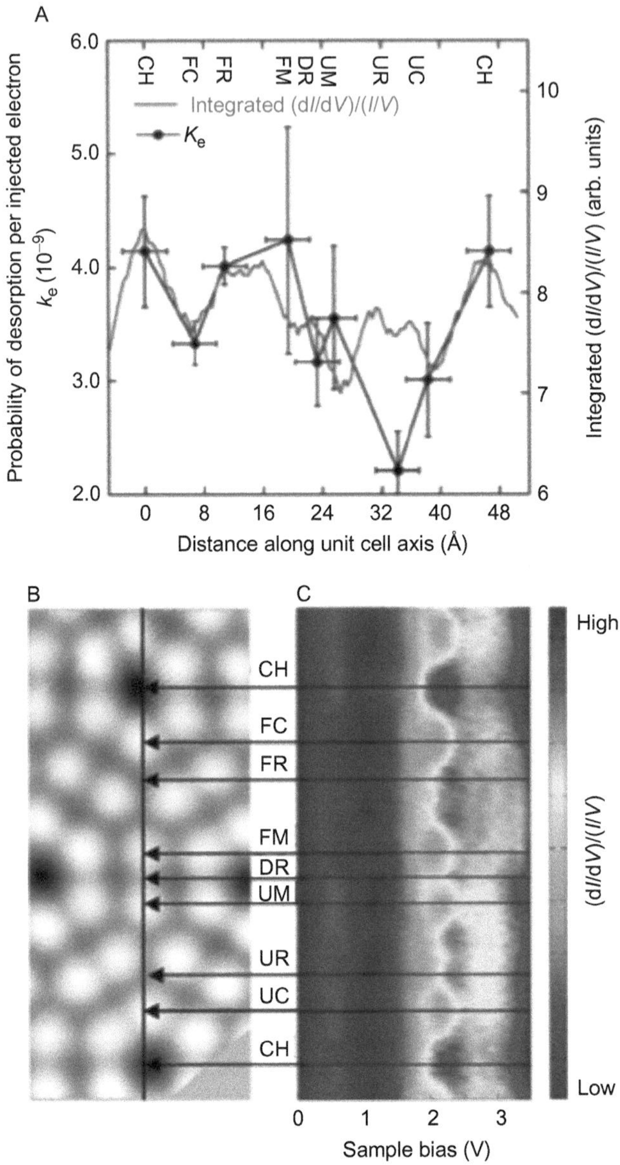

FIGURE 5.19 Site-specific nonlocal chlorobenzene desorption and STS on Si(111)-7 × 7.[99] (A) Probability per electron at +2.7 injection bias of inducing desorption at eight distinct injection sites as indicated in STM image (B): CH, corner hole; FC, faulted corner; FR, faulted restatom; FM, faulted middle; DR, dimer-row; UM, unfaulted middle; UR, unfaulted restatom; and UC, unfaulted corner. (C) Empty states STS map taken along the high-symmetry line joining corner hole to corner hole across a single unit cell. An integrated section of (C) from +2.1 to +2.7 V is plotted for comparison in (A).

REFERENCES

1. Eigler DM, Schweizer EK. Positioning single atoms with a scanning tunneling microscope. *Nature* 1990;**344**:524.
2. Stroscio JA, Eigler DM. Atomic and molecular manipulation with the scanning tunneling microscope. *Science* 1991;**254**:1319.
3. Zeppenfeld P, Lutz CP, Eigler DM. Manipulating atoms and molecules with a scanning tunneling microscope. *Ultramicroscopy* 1992;**42**:128.
4. Crommie MF, Lutz CP, Eigler DM. Confinement of electrons to quantum corrals on a metal-surface. *Science* 1993;**262**:218.
5. Bartels L, Meyer G, Rieder KH. Basic steps involved in the lateral manipulation of single co molecules and rows of co molecules. *Chem Phys Lett* 1997;**273**:371.
6. Bartels L, Meyer G, Rieder KH. Basic steps of lateral manipulation of single atoms and diatomic clusters with a scanning tunneling microscope tip. *Phys Rev Lett* 1997;**79**:697.
7. Meyer G, Bartels L, Zophel S, Henze E, Rieder KH. Controlled atom by atom restructuring of a metal surface with the scanning tunneling microscope. *Phys Rev Lett* 1997;**78**:1512.
8. Hla SW, Kuhnle A, Bartels L, Meyer G, Rieder KH. Controlled lateral manipulation of single diiodobenzene molecules on the Cu(111) surface with the tip of a scanning tunnelling microscope. *Surf Sci* 2000;**454**:1079.
9. Heinrich AJ, Lutz CP, Gupta JA, Eigler DM. Molecule cascades. *Science* 2002;**298**:1381.
10. Keeling DL, Humphry MJ, Fawcett RHJ, Beton PH, Hobbs C, Kantorovich L. Bond breaking coupled with translation in rolling of covalently bound molecules. *Phys Rev Lett* 2005;**94**:4.
11. Grill L, Rieder KH, Moresco F, Rapenne G, Stojkovic S, Bouju X, et al. Rolling a single molecular wheel at the atomic scale. *Nat Nanotechnol* 2007;**2**:95.
12. Whitman LJ, Stroscio JA, Dragoset RA, Celotta RJ. Manipulation of adsorbed atoms and creation of new structures on room-temperature surfaces with a scanning tunneling microscope. *Science* 1991;**251**:1206.
13. Saranin AA, Numata T, Kubo O, Tani H, Katayama M, Lifshits VG, et al. Stm tip-induced diffusion of in atoms on the si(111)root 3x root 3-in surface. *Phys Rev B* 1997;**56**:7449.
14. Mo YW. Reversible rotation of antimony dimers on the silicon (001) surface with a scanning tunneling microscope. *Science* 1993;**261**:886.
15. Ohara M, Kim Y, Kawai M. Electric field response of a vibrationally excited molecule in an Stm junction. *Phys Rev B* 2008;**78**:4.
16. Hosaka S, Hosoki S, Hasegawa T, Koyanagi H, Shintani T, Miyamoto M. Fabrication of nanostructures using scanning probe microscopes. *J Vacuum Sci Technol B* 1995;**13**:2813.
17. Baba M, Matsui S. Atomic desorption of chlorine adsorbed on Si(111) with a scanning tunneling microscope. *Appl Phys Lett* 1994;**64**:2852.
18. Kobayashi A, Grey F, Williams RS, Aono M. Formation of nanometer-scale grooves in silicon with a scanning tunneling microscope. *Science* 1993;**259**:1724.
19. Lyo IW, Avouris P. Field-induced nanometer scale to atomic-scale manipulation of silicon surfaces with the Stm. *Science* 1991;**253**:173.
20. Rezaei MA, Stipe BC, Ho W. Atomically resolved adsorption and scanning tunneling microscope induced desorption on a semiconductor: no on Si(111)-(7 $\times$ 7). *J Chem Phys* 1999;**110**:4891.
21. Avouris P, Walkup RE. Fundamental mechanisms of desorption and fragmentation induced by electronic-transitions at surfaces. *Annu Rev Phys Chem* 1989;**40**:173.
22. Palmer RE. Electron-molecule dynamics at surfaces. *Prog Surf Sci* 1992;**41**:51.

23. Palmer RE, Rous PJ. Resonances in electron-scattering by molecules on surfaces. *Rev Mod Phys* 1992;**64**:383.
24. Stipe BC, Rezaei MA, Ho W, Gao S, Persson M, Lundqvist BI. Single-molecule dissociation by tunneling electrons. *Phys Rev Lett* 1997;**78**:4410.
25. Hla SW, Rieder KH. Stm control of chemical reactions: single-molecule synthesis. *Annu Rev Phys Chem* 2003;**54**:307.
26. Lorente N, Rurali R, Tang H. Single-molecule manipulation and chemistry with the Stm. *J Phys Condens Matter* 2005;**17**:S1049.
27. Hou JG, Zhao AD. Detecting and manipulating single molecules with Stm. *Nano* 2006;**1**:15.
28. Mayne AJ, Dujardin G, Comtet G, Riedel D. Electronic control of single-molecule dynamics. *Chem Rev* 2006;**106**:4355.
29. Otero R, Rosei F, Besenbacher F. Scanning tunneling microscopy manipulation of complex organic molecules on solid surfaces. *Annu Rev Phys Chem* 2006;**57**:497.
30. Comtet G, Dujardin G, Mayne AJ, Riedel D. Principles of operating molecular nanomachines by electronic excitation. *J Phys Condens Matter* 2006;**18**:S1927.
31. Crommie MF, Lutz CP, Eigler DM, Heller EJ. Waves on a metal-surface and quantum corrals. *Surf Rev Lett* 1995;**2**:127.
32. Hla SW, Braun KF, Rieder KH. Single-atom manipulation mechanisms during a quantum corral construction. *Phys Rev B* 2003;**67**:201402(R).
33. Sloan PA. Time-resolved scanning tunnelling microscopy for molecular science. *J Phys Condens Matter* 2010;**22**.
34. Vang RT, Lauritsen JV, Laegsgaard E, Besenbacher F. Scanning tunneling microscopy as a tool to study catalytically relevant model systems. *Chem Soc Rev* 2008;**37**:2191.
35. Kamat PV. Photochemistry on nonreactive and reactive (semiconductor) surfaces. *Chem Rev* 1993;**93**:267.
36. Ageev VN. Desorption induced by electronic-transitions. *Prog Surf Sci* 1994;**47**:55.
37. Madey TE. History of desorption induced by electronic-transitions. *Surf Sci* 1994;**299**:824.
38. Wolkow RA, Moffatt DJ. The frustrated motion of benzene on the surface of Si(111). *J Chem Phys* 1995;**103**:10696.
39. Lo RL, Hwang IS, Ho MS, Tsong TT. Diffusion of single hydrogen atoms on Si(111)-(7 $\times$ 7) surfaces. *Phys Rev Lett* 1998;**80**:5584.
40. Kawasaki T, Sakai D, Kishimoto H, Akbar AA, Ogawa T, Oshima C. Adsorption and desorption of benzene on Si(111)-7 $\times$ 7 studied by scanning tunnelling microscopy. *Surf Interface Anal* 2001;**31**:126.
41. Sakulsermsuk S, Sloan PA, Theis W, Palmer RE. Calibrating thermal and Stm-induced desorption and diffusion for the chemisorbed chlorobenzene/Si(111)7 $\times$ 7 system. *J Phys Condens Matter* 2010;**22**:.
42. Boland JJ. Evidence of pairing and its role in the recombinative desorption of hydrogen from the Si(100)-2 $\times$ 1 surface. *Phys Rev Lett* 1991;**67**:1539.
43. Eigler DM, Lutz CP, Rudge WE. An atomic switch realized with the scanning tunneling microscope. *Nature* 1991;**352**:600.
44. Misewich JA, Heinz TF, Newns DM. Desorption induced by multiple electronic-transitions. *Phys Rev Lett* 1992;**68**:3737.
45. Gao SW, Persson M, Lundqvist BI. Atomic switch proves importance of electron-hole pair mechanism in processes on metal-surfaces. *Solid State Commun* 1992;**84**:271.
46. Saenz JJ, Garcia N. Quantum atom switch—tunneling of Xe atoms. *Phys Rev B* 1993;**47**:7537.

47. Walkup RE, Newns DM, Avouris P. Role of multiple inelastic transitions in atom-transfer with the scanning tunneling microscope. *Phys Rev B* 1993;**48**:1858.
48. Foley ET, Kam AF, Lyding JW, Avouris P. Cryogenic Uhv-Stm study of hydrogen and deuterium desorption from Si(100). *Phys Rev Lett* 1998;**80**:1336.
49. Soukiassian L, Mayne AJ, Carbone M, Dujardin G. Atomic-scale desorption of H atoms from the Si(100)-2x1 : H surface: inelastic electron interactions. *Phys Rev B* 2003;**68**:5.
50. Tong X, Wolkow RA. Electron-induced H atom desorption patterns created with a scanning tunneling microscope: implications for controlled atomic-scale patterning on H-Si(100). *Surf Sci* 2006;**600**:L199.
51. Walsh MA, Hersam MC. Atomic-scale templates patterned by ultrahigh vacuum scanning tunneling microscopy on silicon. *Annu Rev Phys Chem* 2009;**60**:193.
52. Piva PG, DiLabio GA, Pitters JL, Zikovsky J, Rezeq M, Dogel S, et al. Field regulation of single molecule conductivity by a charged surface atom. *Nature* 2005;**435**:658.
53. Saalfrank P. Quantum dynamical approach to ultrafast molecular desorption from surfaces. *Chem Rev* 2006;**106**:4116.
54. Redhead PA. Interaction of slow electrons with chemisorbed oxygen. *Can J Phys* 1964;**42**:886.
55. Antoniewicz PR. Model for electron-stimulated and photon-stimulated desorption. *Phys Rev B* 1980;**21**:3811.
56. Saalfrank P, Boendgen G, Corriol C, Nakajima T. Direct and indirect diet and dimet from semiconductor and metal surfaces: what can we learn from "Toy Models"? (Vol 117, Pg 65, 2000). *Faraday Discuss* 2000;**117**:352.
57. Zewail AH. Femtochemistry: atomic-scale dynamics of the chemical bond. *J Phys Chem A* 2000;**104**:5660.
58. Shen TC, Wang C, Abeln GC, Tucker JR, Lyding JW, Avouris P, et al. Atomic-scale desorption through electronic and vibrational-excitation mechanisms. *Science* 1995;**268**:1590.
59. Shen TC, Steckel JA, Jordan KD. Electron stimulated bond rearrangements on the H/Si (100)-3 × 1 surface. *Surf Sci* 2000;**446**:211.
60. Soukiassian L, Mayne AJ, Carbone M, Dujardin G. Atomic wire fabrication by Stm induced hydrogen desorption. *Surf Sci* 2003;**528**:121.
61. Stokbro K, Thirstrup C, Sakurai M, Quaade U, Hu BYK, Perez-Murano F, et al. Stm-induced hydrogen desorption via a hole resonance. *Phys Rev Lett* 1998;**80**:2618.
62. Avouris P, Walkup RE, Rossi AR, Shen TC, Abeln GC, Tucker JR, et al. Stm-induced H atom desorption from Si(100): isotope effects and site selectivity. *Chem Phys Lett* 1996;**257**:148.
63. Becker RS, Higashi GS, Chabal YJ, Becker AJ. Atomic scale conversion of clean Si(111)-H-1 × 1 to Si(111)-2 × 1 by electron-stimulated desorption. *Phys Rev Lett* 1990;**65**:1917.
64. Aeschlimann , Wolf , Gauyacq , Winter , Shluger , Kleyn , et al. General discussion. *Faraday Discuss* 2000;**117**:257.
65. Mayne AJ, Rose F, Dujardin G. Inelastic interactions of tunnel electrons with surfaces. *Faraday Discuss* 2000;**117**:241.
66. Yuan QH, Li QS, Wang KD, Liu ZF. Theoretical investigation of an intermediate in the Stm tip-induced atomic process on H/Si(100) surfaces. *Phys Rev B* 2010;**81**:7.
67. Bellec A, Riedel D, Dujardin G, Boudrioua O, Chaput L, Stauffer L, et al. Electronic properties of the N-doped hydrogenated silicon (100) surface and dehydrogenated structures at 5 K. *Phys Rev B* 2009;**80**:6.

68. Mayne AJ, Riedel D, Comtet G, Dujardin G. Atomic-scale studies of hydrogenated semiconductor surfaces. *Prog Surf Sci* 2006;**81**:1.
69. Sloan PA, Hedouin MFG, Palmer RE, Persson M. Mechanisms of molecular manipulation with the scanning tunneling microscope at room temperature: chlorobenzene/Si(111)-(7 × 7). *Phys Rev Lett* 2003;**91**:.
70. Sloan PA, Palmer RE. Tip-state control of rates and branching ratios in atomic manipulation. *Nano Lett* 2005;**5**:835.
71. Sloan PA, Palmer RE. Manipulation of polyatomic molecules with the scanning tunnelling microscope at room temperature: chlorobenzene adsorption and desorption from Si(111)-(7 × 7). *J Phys Condens Matter* 2006;**18**:S1873.
72. Lu XK, Polanyi JC, Yang J. A reversible molecular switch based on pattern-change in chlorobenzene and toluene on a Si(111)-(7 × 7) surface. *Nano Lett* 2006;**6**:809.
73. Lu PH, Polanyi JC, Rogers D. Electron-induced "localized atomic reaction" (Lar): chlorobenzene adsorbed on Si(111) 7 × 7. *J Chem Phys* 1999;**111**:9905.
74. Palmer RE, Sloan PA, Xirouchaki C. Decoration of surfaces with size-selected clusters and molecular manipulation at room temperature: precision and uncertainty in organizing atoms. *Philos Trans R Soc A* 2004;**362**:1195.
75. Sloan PA, Palmer RE. Two-electron dissociation of single molecules by atomic manipulation at room temperature. *Nature* 2005;**434**:367.
76. Cao Y, Deng JF, Xu GQ. Stereo-selective binding of chlorobenzene on Si(111)-7 × 7. *J Chem Phys* 2000;**112**:4759.
77. Piancastelli MN, Motta N, Sgarlata A, Balzarotti A, Decrescenzi M. Topographic and spectroscopic analysis of ethylene adsorption on Si(111)7 × 7 by Stm and Sts. *Phys Rev B* 1993;**48**:17892.
78. Shachal D, Manassen Y, TerOvanesyan E. Role of chain length on the surface chemistry of Dienes studied by scanning tunneling microscopy. *Phys Rev B* 1997;**55**:9367.
79. Miyake K, Ishida M, Shigekawa H. Selective chemical reaction of Hbo2 molecules on the Si (111)-7 × 7 surface studied by scanning tunneling microscopy. *Appl Surf Sci* 1998;**130**:78.
80. Fukuda Y, Shimomura M, Kaneda G, Sanada N, Zavodinsky VG, Kuyanov IA, et al. Scanning tunneling microscopy, high-resolution electron energy loss spectroscopy, and theoretical studies of trimethylphosphine (Tmp) on a Si(111)-(7 × 7) surface. *Surf Sci* 1999;**442**:507.
81. Frederick BG, Munz A, Bertrams T, Chen Q, Perry CC, Richardson NV. Negative ion mediated molecular manipulation with Stm of C(8 × 2) benzoate/Cu(110). *Chem Phys Lett* 1997;**272**:438.
82. Dujardin G, Mayne A, Robert O, Rose F, Joachim C, Tang H. Vertical manipulation of individual atoms by a direct Stm tip-surface contact on Ge(111). *Phys Rev Lett* 1998;**80**:3085.
83. Brown DE, Moffatt DJ, Wolkow RA. Isolation of an intrinsic precursor to molecular chemisorption. *Science* 1998;**279**:542.
84. Wolkow RA. Controlled molecular adsorption on silicon: laying a foundation for molecular devices. *Annu Rev Phys Chem* 1999;**50**:413.
85. Alavi S, Rousseau R, Lopinski GP, Wolkow RA, Seideman T. Controlling organic reactions on silicon surfaces with a scanning tunneling microscope: theoretical and experimental studies of resonance-mediated desorption. *Faraday Discuss* 2000;**117**:213.
86. Alavi S, Rousseau R, Patitsas SN, Lopinski GP, Wolkow RA, Seideman T. Inducing desorption of organic molecules with a scanning tunneling microscope: theory and experiments. *Phys Rev Lett* 2000;**85**:5372.
87. Hofer WA, Fisher AJ, Lopinski GP, Wolkow RA. Adsorption of benzene on Si(100)-(2 × 1): adsorption energies and Stm image analysis by ab initio methods. *Phys Rev B* 2001;**63**:7.

88. Kruse P, Wolkow RA. "Gentle lithography" with benzene on Si(100). *Appl Phys Lett* 2002;**81**:4422.
89. Patitsas SN, Lopinski GP, Hul'ko O, Moffatt DJ, Wolkow RA. Current-induced organic molecule-silicon bond breaking: consequences for molecular devices. *Surf Sci* 2000;**457**: L425.
90. Hla SW, Bartels L, Meyer G, Rieder KH. Inducing all steps of a chemical reaction with the scanning tunneling microscope tip: towards single molecule engineering. *Phys Rev Lett* 2000;**85**:2777.
91. Ullmann F, Meyer G, Loewenthal O, Gilli E. Symmetric biphenyl derivates. *Justus Liebigs Annalen Der Chemie* 1904;**332**:38.
92. Salam GP, Persson M, Palmer RE. Possibility of coherent multiple excitation in atom-transfer with a scanning tunneling microscope. *Phys Rev B* 1994;**49**:10655.
93. Harikumar KR, Leung L, McNab IR, Polanyi JC, Lin HP, Hofer WA. Cooperative molecular dynamics in surface reactions. *Nat Chem* 2009;**1**:716.
94. Stipe BC, Rezaei MA, Ho W. A variable-temperature scanning tunneling microscope capable of single-molecule vibrational spectroscopy. *Rev Sci Instrum* 1999;**70**:137.
95. Stipe BC, Rezaei MA, Ho W. Inducing and viewing the rotational motion of a single molecule. *Science* 1998;**279**:1907.
96. Jiang GP, Polanyi JC, Rogers D. Electron and photon irradiation of benzene and chlorobenzene on Si(111) 7 × 7. *Surf Sci* 2003;**544**:147.
97. Dressler R, Allan M, Haselbach E. Symmetry control in bond-cleavage processes—dissociative electron-attachment to unsaturated halocarbons. *Chimia* 1985;**39**:385.
98. Yoder NL, Guisinger NP, Hersam MC, Jorn R, Kaun CC, Seideman T. Quantifying desorption of saturated hydrocarbons from silicon with quantum calculations and scanning tunneling microscopy. *Phys Rev Lett* 2006;**97**:187601.
99. Sloan PA, Sakulsermsuk S, Palmer RE. Nonlocal desorption of chlorobenzene molecules from the Si(111)-(7 × 7) surface by charge injection from the tip of a scanning tunneling microscope: remote control of atomic manipulation. *Phys Rev Lett* 2010;**105**:048301.
100. McNab IR, Polanyi JC. Patterned atomic reaction at surfaces. *Chem Rev* 2006;**106**:4321.
101. Braunschweig AB, Huo FW, Mirkin CA. Molecular printing. *Nat Chem* 2009;**1**:353.
102. Harikumar KR, Lim TB, McNab IR, Polanyi JC, Zotti L, Ayissi S, et al. Dipole-directed assembly of lines of 1,5-dichloropentane on silicon substrates by displacement of surface charge. *Nat Nanotechnol* 2008;**3**:222.
103. Dobrin S, Harikumar KR, Jones RV, Li N, McNab IR, Polanyi JC, et al. Self-assembled molecular corrals on a semiconductor surface. *Surf Sci* 2006;**600**:L43.
104. Harikumar KR, Polanyi JC, Sloan PA, Ayissi S, Hofer WA. Electronic switching of single silicon atoms by molecular field effects. *J Am Chem Soc* 2006;**128**:16791.
105. Maksymovych P, Sorescu DC, Jordan KD, Yates JT. Collective reactivity of molecular chains self-assembled on a surface. *Science* 2008;**322**:1664.
106. Lopinski GP, Wayner DDM, Wolkow RA. Self-directed growth of molecular nanostructures on silicon. *Nature* 2000;**406**:48.
107. Ho W. Single-molecule chemistry. *J Chem Phys* 2002;**117**:11033.
108. Morgenstern K, Rieder KH. Dissociation of water molecules with the scanning tunnelling microscope. *Chem Phys Lett* 2002;**358**:250.
109. Ueba H. Motions and reactions of single adsorbed molecules induced by vibrational excitation with Stm. *Surf Rev Lett* 2003;**10**:771.
110. Repp J, Meyer G, Paavilainen S, Olsson FE, Persson M. Imaging bond formation between a gold atom and pentacene on an insulating surface. *Science* 2006;**312**:1196.

111. Katano S, Kim Y, Hori M, Trenary M, Kawai M. Reversible control of hydrogenation of a single molecule. *Science* 2007;**316**:1883.
112. Guo H, Ji W, Polanyi JC, Yang JSY. Molecular dynamics of localized reaction, experiment and theory: methyl bromide on Si(111)-7 × 7. *ACS Nano* 2008;**2**:699.
113. Matta CF, Polanyi JC. Chemistry on a peg-board: the effect of adatom-adatom separation on the reactivity of dihalobenzenes at Si(111)7 × 7 surfaces. *Philos Trans R Soc A* 2004;**362**:1185.
114. Repp J, Meyer G, Olsson FE, Persson M. Controlling the charge state of individual gold adatoms. *Science* 2004;**305**:493.
115. Liljeroth P, Repp J, Meyer G. Current-induced hydrogen tautomerization and conductance switching of naphthalocyanine molecules. *Science* 2007;**317**:1203.
116. Gross L, Mohn F, Liljeroth P, Repp J, Giessibl FJ, Meyer G. Measuring the charge state of an adatom with noncontact atomic force microscopy. *Science* 2009;**324**:1428.
117. Gawronski H, Carrasco J, Michaelides A, Morgenstern K. Manipulation and control of hydrogen bond dynamics in absorbed ice nanoclusters. *Phys Rev Lett* 2008;**101**:4.
118. Sicot M, Kurnosikov O, Adam OAO, Swagten HJM, Koopmans B. Stm-induced desorption of hydrogen from Co nanoislands. *Phys Rev B* 2008;**77**:035417.
119. Chen L, Li H, Wee ATS. Nonlocal chemical reactivity at organic-metal interfaces. *ACS Nano* 2009;**3**:3684.
120. Nouchi R, Masunari K, Ohta T, Kubozono Y, Iwasa Y. Ring of C-60 polymers formed by electron or hole injection from a scanning tunneling microscope tip. *Phys Rev Lett* 2006;**97**.
121. Nakamura Y, Mera Y, Maeda K. Diffusion of chlorine atoms on Si(111)-(7 × 7) surface enhanced by electron injection from scanning tunneling microscope tips. *Surf Sci* 2001;**487**:127.
122. Nakamura Y, Mera Y, Maeda K. Nanoscale imaging of electronic surface transport probed by atom movements induced by scanning tunneling microscope current. *Phys Rev Lett* 2002;**89**:266805.
123. Maeda K, Nakamura Y. Spreading effects in surface reactions induced by tunneling current injection from an Stm tip. *Surf Sci* 2003;**528**:110.
124. Nakamura Y, Mera Y, Maeda K. Hopping motion of chlorine atoms on Si(100)-(2 × 1) surfaces induced by carrier injection from scanning tunneling microscope tips. *Surf Sci* 2003;**531**:68.
125. Nakamura Y, Mera Y, Maeda K. Desorption of chlorine atoms on Si(111)-(7 × 7) surfaces induced by hole injection from scanning tunneling microscope tips. *Surf Sci* 2007;**601**:2189.
126. MacLeod JM, Lipton-Duffin J, Fu CY, Rosei F. Inducing nonlocal reactions with a local probe. *ACS Nano* 2009;**3**:3347.
127. Maksymovych P, Dougherty DB, Zhu XY, Yates JT. Nonlocal dissociative chemistry of adsorbed molecules induced by localized electron injection into metal surfaces. *Phys Rev Lett* 2007;**99**:016101.
128. Bellec A, Riedel D, Dujardin G, Boudrioua O, Chaput L, Stauffer L, et al. Nonlocal activation of a bistable atom through a surface state charge-transfer process on Si(100)-(2 × 1):H. *Phys Rev Lett* 2010;**105**:048302.
129. Tomatsu K, Nakatsuji K, Yamada M, Komori F, Yan BH, Yam CY, et al. Local vibrational excitation through extended electronic states at a germanium surface. *Phys Rev Lett* 2009;**103**:266102.
130. Sloan PA, Sakulsermsuk S, Palmer RE. To be published.

Index

Note: The letters '*f*' and '*t*' following the locators refer to figures and tables respectively.

A

Adsorbate-substrate interactions, 17–18
AFM. *See* Atomic force microscopy
Ag adatom on NaCal/Cu
  Au adatom on NaCal/Cu compared to, 27–28
  charge state assignment on, 28
  IS and, 28, 29*f*
  STM imaging different states of, 29*f*
Alkali halide films, 18–19
Ammonia, 79–81
Anderson impurity model, 67–68
Antoniewicz model, 124, 125*f*
Atomic and molecular manipulation, 1. *See also* Direct contact method, STM; Inelastic electron tunnelling, STM; Scanning tunnelling microscope; STM tunnel current-induced manipulation
  areas of, 121
  of Au adatoms on NaCal/Cu, 24–25, 25*f*
    quantum yield as function of voltage in, 25–27, 26*f*
  developments in, 1–2
  experimental methods of, 122–123
  on metal surfaces, 2, 4*f*
  molecular orbital engineering and, for small molecules, 36
  for molecule-metal-molecule bridge, 41
  neutral-to-negative charge state process of, 25–26
  nonlocal, 138–144
    of CH_3SSCH_3 molecule at Au(111), 139–141, 140*f*
    of chlorobenzene at Si(111)-7 x 7, 141–144, 142*f*, 144*f*
  progress and developments in, 9–10
  quantum mirage experiment in, 2, 3*f*
  techniques for, 17–18
  of ultra-thin insulating films, 44–45
  volumetric quantum holography and, 2, 4*f*
Atomic force microscopy (AFM)
  lateral manipulation of, 7
  of pentacene molecule, 7–9, 9*f*
  versatility of, 7–9
  vertical manipulation of, 7, 8*f*
Atomically precise nanostructures, 17–18
Au
  TCNQ on, 54–57
    differential conductance spectra on, 56–57, 56*f*
    hydrogen bond network in, 55–56
    molecular structure of, 55, 55*f*
  TTF on, 57–65
    DFT of, 62, 63–64, 63*f*
    disordered distribution of, 57–58, 58*f*
    intermolecular distance of, 58–59
    labyrinthine patten of, 64–65
    mean interaction potentials of one-dimensional, 60, 61*f*
    nucleation of, 64–65, 64*f*
    pair distributions of, 59–60, 59*f*
    STM images for, 58*f*, 64*f*
  TTF-TCNQ compound on, 65–70
    Kondo resonance and, 65, 67, 69
    STM images for, 66*f*
    ZBP in, 65
Au(111), nonlocal manipulation of CH_3SSCH_3 at, 139–141, 140*f*
Au adatoms on NaCal/Cu, 24–28
  Ag adatom on NaCal/Cu compared to, 27–28
  DFT simulations for, 25
  diffusion experiments for, 27
  manipulation of, 24–25, 25*f*
    quantum yield as function of voltage in, 25–27, 26*f*
  NIR state of, 25–26, 27
  scattering and, 24, 25*f*
Au-pentacene
  bond formation of, 36–37, 38*f*
  charge state of, 38
  DFT for, 39–40
  d*I*/d*V* of, 38, 39*f*
  HOMO, LUMO, SOMO of, 39–40
  IET on insulating films and, 36–37
  organometallic synthesis of, 36–40

Aviram, A, 51–53
Avouris, P, 79–81

B

Benzene
physisorbed circles of, 99–102, 101*f*, 102*f*
on Si(100), DIET of, 128–130, 130*f*
Bias-dependent imaging, 30–31
Boltzmann factor, 60
Bond dissociation. *See* STM-induced bond dissociation
Bulk insulators, 17–18

C

CB. *See* Coulomb Blockade
CdSe nanorods, 7–9
CDW. *See* Charge density waves
CH. *See* Corner hole
CH_3SSCH_3 molecule, nonlocal manipulation of, at Au(111), 139–141, 140*f*
Chain reaction, 108–109, 109*f*, 110*f*
Charge density waves (CDW), 53
Charge distribution, Smoluchowski effect of metal surfaces and, 20–22
Charge state
Ag adatom on NaCal/Cu assignment of, 28
of Au-pentacene, 38
IS determining, 22–23
manipulation and neutral-to-negative, 25–26
manipulation of metal atoms, controlling, 24–28
Charge transfer complexes. *See* Organic charge transfer complexes
Chemical reactions. *See* Reaction, modes of
Chemistry at atomic scale, 1
Chlorine, dissociative attachment of, at Si(100)-2 x 1, 86
Chlorobenzene
gas phase dissociative electron attachment of, 135–137
at Si(111)-7 x 7, 81–82, 82*f*
desorption of, 127–131, 127*f*, 128*f*, 129*f*
DFT for, 82–83, 83*f*
electron-induced reaction of, 82
nonlocal manipulation of, 141–144, 142*f*, 144*f*
STM-induced bond dissociation of, 134–137, 135*f*, 136*f*
Circle imprinting, 99–102, 101*f*, 102*f*
Co islands on Cu(111), nonlocal manipulation of, 138–139, 139*f*
Constant orbital probability distribution contours, 32*f*
Cooperative reaction, 107, 108*f*
Coordination chemistry, 40
Coordination complex synthesis, 5–6
Corner hole (CH), 143
Coulomb Blockade (CB), 67
CT complexes. *See* Organic charge transfer complexes
CT solids, 51–52
Cu. *See also* Ag adatom on NaCal/Cu; Au adatoms on NaCal/Cu
NaCl films and
Ag adatoms on, 27–28
Au adatoms on, 24–28
binding geometry for, 20–22, 21*f*
growth of, 19–22, 20*f*
intrinsically stepped, 20–22
IS in, 22–24, 23*f*
Cu substrates
growth behavior of, 19–20
intrinsically stepped, 20–22
NaCl films and
binding geometry for, 20–22, 21*f*
IS in, 22–24, 23*f*
CuN films, monatomic metal wires on, 28–30, 30*f*

D

DDA. *See* Dipole-directed assembly
Density-functional theory (DFT)
for Au adatoms on NaCal/Cu, 25
Au-pentacene and, 39–40
of chlorobenzene at Si(111)-7 x 7, 82–83, 83*f*
NaCl films and, 19–20
of TTF on Au, 62, 63–64, 63*f*
Desorption
of chlorobenzene at Si(111)-7 x 7, 127–131, 127*f*, 128*f*, 129*f*
electric field affecting, 128–130
of hydrogen from Si(100)-2 x 1, 126–127, 126*f*
STM and, 123–131
STM tunnel current-induced manipulation and, 128–130
Desorption induced by electronic transition (DIET), 124–125
of benzene on Si(100), 128–130, 130*f*
DIMET crossover between, 125
models of, 124, 125*f*
Desorption induced by multiple electronic transitions (DIMET), 123, 124–125
DIET crossover between, 125

DFT. *See* Density-functional theory
Diatomic molecules, dissociative attachment for, at Si(100)-2 x 1, 84–86, 85*f*
  of chlorine, 86
  of H_2, 84–86
  of oxygen, 86
Dibromobenzenes, double dissociative attachment, 92, 94*f*
  at Si(111)-7 x 7, 92–94, 95*f*, 96*f*
Dichloropentane, 103
d*I*/d*V*. *See* Differential conductance
DIET. *See* Desorption induced by electronic transition
Differential conductance (d*I*/d*V*), 22, 29*f*, 37–38
  of Au-pentacene, 38, 39*f*
  in TCNQ, 69–70, 71*f*
DIMET. *See* Desorption induced by multiple electronic transitions
Dipole-directed assembly (DDA), 103
Dipoles, TTF interactions and, 61–62
Direct contact method, STM, 2
Direct reaction, 106–107
Dissociative attachment
  of diatomic molecules at Si(100)-2 x 1, 84–86, 85*f*
    of chlorine, 86
    of H_2, 84–86
    of oxygen, 86
  double
    of dibromobenzenes, 92, 94*f*
    of dibromobenzenes at Si(111)-7 x 7, 92–94, 95*f*, 96*f*
  energy barrier to, 91, 91*f*
  general predictive guides for, 91
  minimum energy paths for, 93*f*
  multiple, 91–94
  of polyatomic molecules at Si(100)-2 x 1, 86–91
    of ammonia, 87
    of methyl bromide, 88–90, 89*f*, 90*f*
    of methyl chloride, 90–91
    of water, 87–88, 88*f*
  single
    diatomic molecules, 84–86
    discussion of, 91
    polyatomic molecules, 86–91
Double-barrier tunnelling junction, 17–18, 18*f*

E

EAs. *See* Electron affinities
Eigler, D. M, 2
Electron affinities (EAs), 51–53
Electron transport, 42–44
Electronics at atomic scale, 1
Electron-phonon coupling, 70
Electrostatic potential, 22–23
Energy transport, 42–44
Ethylene, 111
  electron-induced migration of, 111*f*
    angular distributions of, 112*f*
  rotational recoil of, 112

F

Fano lineshape, 67
Fe(DCA)$_2$ complex, 40–41
Fluoropentane, STM induced bond dissociation of, at Si(100), 137–138, 138*f*
Forming lines, 103, 104*f*
Frame-by-frame scanning, 122
Franck-Condon principle, 34–35, 34*f*
Friedel oscillations, 62
Full width at half maximum (FWHM), 34–35

H

H_2, dissociative attachment of, at Si(100)-2 x 1, 84–86
Haloalkane molecules at Si(100)-2 x 1, 103
Heterogeneous catalysis, 79
Highest occupied molecular orbital (HOMO), 30–31, 32*f*
  of Au-pentacene, 39–40
  HOMO-LUMO peak separation, 38
  STM of naphthalocyanine molecule at, 41–42, 42*f*
HOMO-LUMO peak separation, 38
Hydrogen
  desorption from Si(100)-2 x 1, 126–127, 126*f*
  tautomerization reaction, 41–44

I

IET. *See* Inelastic electron tunnelling
IETS. *See* Inelastic electron tunnelling spectroscopy
Imprinting. *See* Pattern imprinting; Single-molecule imprinting
Indirect reaction, 106–107
Inelastic electron tunnelling (IET)
  on insulating films, 35
    Au-pentacene and, 36–37
  of naphthalocyanine molecule, 42, 43*f*
  STM using, 2–5, 35
  types of, 35
Inelastic electron tunnelling spectroscopy (IETS), 36–37

Insulating films. *See also* Ag adatom on NaCal/Cu; Au adatoms on NaCal/Cu
  alkali halide, 18–19
  CuN, monatomic metal wires on, 28–30, 30*f*
  IET on, 35–36
    Au-pentacene and, 36–37
  LDOS reduced by, 35
  MgO, water dissociation on, 36
  NaCl, 19
  NaCl with Cu and Cu substrates
    binding geometry for, 20–22, 21*f*
    DFT simulations with, 19–20
    growth of, 19–22, 20*f*
    intrinsically stepped, 20–22
    IS in, 22–24, 23*f*
    IS probability distribution for, 22–23, 23*f*
    thickness of, 20*f*
  oxide, 18–19
  pentacene molecules life extended with, 33
  STM and, 17–18
  ultra-thin, 17–22
    atomic and molecular manipulation of, 44–45
    TTF-TCNQ compound and, 53
Interface state (IS)
  Ag adatoms on NaCl/Cu and, 28, 29*f*
  charge state determined with, 22–23
  in NaCl/Cu, 22–24, 23*f*
    probability distribution for, 22–23, 23*f*
Intermolecular energy, 42–44
Intramolecular bond dissociation, 131–138
Ionization energy (IP), 51–53
IP. *See* Ionization energy
IS. *See* Interface state

K

Kondo resonance, 6
  inelastic, mediated by vibron-assisted tunnelling, 69–70
  TTF-TCNQ compound on Au and, 65, 67, 69
  ZBP and, 68–70, 68*f*
Kondo-Abrikosov-Suhl resonance, 67

L

LAR. *See* Localized atomic reactions
Lateral manipulation with metal surfaces, 5–6
LDOS. *See* Local density of states
Lines
  forming, 103, 104*f*
  MSI of, 103–106, 105*f*
Local density of states (LDOS), 35, 143
  insulating film reducing, 35–36
Localized atomic reactions (LAR), 6
  of chlorobenzene at Si(111)-7 x 7, 81–82, 82*f*
    DFT for, 82–83, 83*f*
    electron-induced reaction of, 82
  on metals, 94–98, 97*f*, 98*f*
    electron-induced reaction in, 95–96, 97–98
    thermal dissociation in, 95, 96, 97*f*
  MSI and, 83–84
  single-molecule imprinting and, 81–84
Lowest unoccupied molecular orbital (LUMO), 30–31, 32*f*
  of Au-pentacene, 39–40
  HOMO-LUMO peak separation, 38
  STM of naphthalocyanine molecule at, 41–42, 42*f*

M

Manipulation of metal atoms
  charge state control for, 24–28
  vertical transfer and build-up of nanostructures in, 28–30
Metal
  LAR on, 94–98, 97*f*, 98*f*
    electron-induced reaction in, 95–96, 97–98
    thermal dissociation in, 95, 96, 97*f*
  single-molecule imprinting on, 94–98
  surfaces
    corrugation of, 20–22
    lateral manipulation with, 5–6
    Smoluchowski effect of charge distribution at, 20–22
Metal atoms, manipulation of, 24–30
  charge state control for, 24–28
  vertical transfer and build-up of nanostructures in, 28–30
Metal-free molecular magnets, 6
Metal-ligand complex formation, 40–41
  STM and STS for, 40
Methyl bromide, dissociative attachment of, at Si(100)-2 x 1, 88–90, 89*f*, 90*f*
Methyl chloride, dissociative attachment of, at Si(100)-2 x 1, 90–91
MgO films, water dissociation on, 36
MGR-A model, 124, 125*f*
MGR-B model, 124, 125*f*
Modes of reaction, 106–113
  chain, 108–109, 109*f*, 110*f*
  cooperative, 107, 108*f*
  direct and indirect, 106–107

localization of, 113–114
recoil, 109–113, 111*f*, 112*f*, 113*f*
Modified image potential, 22–23
Molecular electronics, 51–53
Molecular magnetism
CT complexes inducing, 65–70
metal-free organization of, 72
Molecular orbital engineering, 5–6, 35–44
basis of, 30
manipulation of small molecules in, 36
metal-ligand complex formation and, 40–41
molecular switching based on tautomerization reaction and, 41–44
organometallic synthesis of Au-pentacene and, 36–40
Molecular scale imprinting (MSI), 6
of lines, 103–106, 105*f*
Molecular switching, tautomerization reaction and, 41–44
Molecular-scale imprinting (MSI)
induced reaction in, 83–84
LAR and, 83–84
Molecule-metal-molecule bridge, 41
Molecules. *See* Atomic and molecular manipulation
Monatomic metal wires, on CuN films, 28–30, 30*f*
Monopoles, TTF as
freezing temperature and, 62
interactions of, 61–62
origin of, 62
MSI. *See* Molecular scale imprinting; Molecular-scale imprinting
Multiple dissociative attachment, 91–94
Multiplexing
pattern imprinting and, 98–99
self-assembly and, 99

N

NaCl films, 19. *See also* Ag adatom on NaCal/Cu; Au adatoms on NaCal/Cu
CU and CU substrates and
Ag adatoms on, 27–28
Au adatoms on, 24–28
binding geometry for, 20–22, 21*f*
growth of, 19–22, 20*f*
intrinsically stepped, 20–22
IS in, 22–24, 23*f*
DFT simulations with, 19–20
electronic decoupling of, 33
thickness of, 20*f*
Nanoelectronics, 5
development of, 10
Franck-Condon principle and, 34–35, 34*f*
Nanoscience
pre-designed patterns at atomic scale in, 137
public awareness of, 1–2
Naphthalocyanine molecule
hydrogen atoms in, 41
IET of, 42, 43*f*
STM images at HOMO, LUMO of, 41–42, 42*f*
NC-AFM, 7–9
NEB. *See* Nudged elastic band
Negative ion-resonance state (NIR state), 25–26, 27
NiAl substrate, 18–19
NIR state. *See* Negative ion-resonance state
Nonlocal manipulation, 138–144
of CH_3SSCH_3 molecule at Au(111), 139–141, 140*f*
of chlorobenzene at Si(111)-7 x 7, 141–144, 142*f*, 144*f*
of Co islands on Cu(111), 138–139, 139*f*
Nudged elastic band (NEB), 100–102

O

Orbital imaging. *See also* Molecular orbital engineering
constant orbital probability distribution contours in, 32*f*
Franck-Condon principle and, 34–35, 34*f*
prerequisites of, 30
Organic charge transfer complexes (CT complexes)
CT solids, 51–52
history and uses of, 51–53
molecular magnetism induced by, 65–70
TCNQ on Au, 54–57
TTF on Au, 57–65
nucleation of, 64–65, 64*f*
TTF-TCNQ compound, 52–53
on Au, 65–70
metallicity of, 52–53
temperature and structural transformations of, 53
ultra-thin film and, 53
Organometallic synthesis, 5–6
of Au-pentacene, 36–40
Oxide films, 18–19
Oxygen
dissociative attachment of, at Si(100)-2 x 1, 86
STM-induced bond dissociation at Pt(111), 132–134, 133*f*, 134*f*

P

Pattern imprinting
- circles, 99–102, 101*f*, 102*f*
- lines
  - forming, 103, 104*f*
  - MSI of, 103–106, 105*f*
- multiplexing and, 98–99
- self-assembly and, 98–106

Pentacene molecules. *See also* Au-pentacene
- AFM imaging of, 7–9, 9*f*
- constant orbital probability distribution contours of, 32*f*
- experiment suitability of, 30
- Franck-Condon principle and, 34–35, 34*f*
- FWHM of, 34–35
- insulating films extending life of, 33
- STS on, 30–31, 32*f*

Pentacene-terminated tip, 30–31

PES. *See* Potential-energy surface

Phase-accumulation model, 22–23, 23*f*

Physisorbed circle imprinting, 99–102, 101*f*, 102*f*

Polyatomic molecules, dissociative attachment of, at Si(100)-2 x 1, 86–91
- of ammonia, 87
- of methyl bromide, 88–90, 89*f*, 90*f*
- of methyl chloride, 90–91
- of water, 87–88, 88*f*

Potential-energy surface (PES), 91
- calculation of, 91, 91*f*

Printing. *See* Single-molecule imprinting

Probability distribution, for IS in NaCl/Cu, 22–23, 23*f*

PT(111), STM-induced bond dissociation of oxygen at, 132–134, 133*f*, 134*f*

Q

Quantum mirage experiment, 2, 3*f*

Quantum yield, 25–27, 26*f*

R

Ratner, M. A, 51–53

Reaction, modes of, 106–113
- chain, 108–109, 109*f*, 110*f*
- cooperative, 107, 108*f*
- direct and indirect, 106–107
- localization of, 113–114
- recoil, 109–113, 111*f*, 112*f*, 113*f*
- Ulmann, 131, 131*f*

Recoil reaction, 109–113, 111*f*, 112*f*, 113*f*

Rotational recoil, 112

S

Scanning tunnelling microscope (STM), 1. *See also* Atomic and molecular manipulation; Inelastic electron tunnelling; STM tunnel current-induced manipulation; STM-induced bond dissociation
- Ag adatom on NaCal/Cu states imaged by, 29*f*
- calculation methods for, 5
- design and operation of, 2
- desorption and, 123–131
- direct contact method for, 2
- double-barrier tunnelling junction and, 17–18, 18*f*
- enhanced imaging with tip of, 30–31
- experimental methods of, 122–123
  - frame-by-frame scanning, 122
  - single-point current injection, 122–123
- IET for, 2–5, 35
- insulating films and, 17–18
- for metal-ligand complex formation, 40
- of naphthalocyanine molecule at HOMO, LUMO, 41–42, 42*f*
- single atom removal and, 7
- single-molecular orbitals imaged with, 5–6
- surface science and, 79–81
- temperature developments with, 9–10
- for TTF on Au, 58*f*, 64*f*
- for TTF-TCNQ compound on Au, 66*f*
- water dissociation induced by, 36

Scanning tunnelling spectroscopy (STS)
- for metal-ligand complex formation, 40
- on pentacene molecules, 30–31, 32*f*

Schrödinger equation, 22–23

Self-assembly, 137. *See also* Reaction, modes of
- multiplexing and, 99
- pattern imprinting and, 98–106

Semiconductors, single-molecule imprinting on, 79–94

Si(100)
- DIET of benzene on, 128–130, 130*f*
- STM-induced bond dissociation of fluoropentane at, 137–138, 138*f*

Si(100)-2 x 1
- dissociative attachment at
  - of ammonia, 87
  - of chlorine, 86
  - of diatomic molecules, 84–86, 85*f*
  - of H_2, 84–86
  - of methyl bromide, 88–90, 89*f*, 90*f*
  - of methyl chloride, 90–91

of oxygen, 86
of polyatomic molecules, 86–91
of water, 87–88, 88*f*
haloalkane molecules at, 103
hydrogen desorption from, 126–127, 126*f*
Si(111)-7 x 7
chlorobenzene at, 81–82, 82*f*
desorption of, 127–131, 127*f*, 128*f*, 129*f*
DFT for, 82–83, 83*f*
electron-induced reaction of, 82
nonlocal manipulation of, 141–144, 142*f*, 144*f*
STM-induced bond dissociation of, 134–137, 135*f*, 136*f*
double dissociative attachment of dibromobenzenes at, 92–94, 95*f*, 96*f*
reconstruction of, 79–81, 80*f*
Silicon carbide, 2–5
Single atom removal, 7
Single dissociative attachment. *See also* Dissociative attachment
diatomic molecules, 84–86
discussion of, 91
polyatomic molecules, 86–91
Single-molecular orbitals, 5–6
Single-molecule chemistry, 17–18
Single-molecule imprinting. *See also* Pattern imprinting
LAR and, 81–84
on metals, 94–98
LAR, 94–98, 97*f*, 98*f*
on semiconductors, 79–94
Single-point current injection, 122–123
Singly occupied molecular orbital (SOMO), 38
of Au-pentacene, 39–40
Smoluchowski, R, 20–22
Smoluchowski effect, 20–22
SOMO. *See* Singly occupied molecular orbital
STM. *See* Scanning tunnelling microscope
STM tunnel current-induced manipulation, 121–122
desorption and, 128–130
STM-induced bond dissociation, 131–138, 131*f*
chlorobenzene at Si(111)-7 x 7, 134–137, 135*f*, 136*f*
fluoropentane at Si(100), 137–138, 138*f*
oxygen at Pt(111), 132–134, 133*f*, 134*f*
STS. *See* Scanning tunnelling spectroscopy
Surface diffusion, 2–5
Surface science, 79–81

T

Tautomerization reaction, molecular switching based on, 41–44
TCNQ. *See* Tetracyanoquinodimethane
Temperature
STM developments with, 9–10
TTF monomers and freezing, 62
TTF-TCNQ compound structural transformations and, 53
Tetracyanoquinodimethane (TCNQ), 6, 51–53. *See also* TTF-TCNQ compound
on Au, 54–57
differential conductance spectra on, 56–57, 56*f*
hydrogen bond network in, 55–56
molecular structure of, 55, 55*f*
d*I*/d*V* in, 69–70, 71*f*
molecular model of, 54*f*
Tetrathiafulvalene (TTF), 6, 51–53. *See also* TTF-TCNQ compound
on Au, 57–65
DFT of, 62, 63–64, 63*f*
disordered distribution of, 57–58, 58*f*
intermolecular distance of, 58–59
labyrinthine patten of, 64–65
mean interaction potentials of one-dimensional, 60, 61*f*
nucleation of, 64–65, 64*f*
pair distributions of, 59–60, 59*f*
STM images for, 58*f*, 64*f*
dipolar interactions between, 61–62
as monopoles
freezing temperature and, 62
interactions of, 61–62
origin of, 62
Thermal dissociation, in LAR on metal, 95, 96, 97*f*
Thermal fluorination, 107, 108*f*
Transition state (TS), 81–82
TTF. *See* Tetrathiafulvalene
TTF-TCNQ compound, 52–53
on Au, 65–70
Kondo resonance and, 65, 67, 69
STM images for, 66*f*
ZBP in, 65
metallicity of, 52–53
temperature and structural transformations of, 53
ultra-thin film and, 53

Tunnel current-induced manipulation. *See* STM tunnel current-induced manipulation

U

UHV. *See* Ultra-high vacuum
Ulmann reaction, steps of, 131, 131*f*
Ultra-high vacuum (UHV), 87–88
Ultra-thin insulating films, 17–22
  atomic and molecular manipulation of, 44–45
  TTF-TCNQ compound and, 53
Unfaulted restatom (UR), 143

V

Volumetric quantum holography, 2, 4*f*

W

Water
  dissociative attachment of, at Si(100)-2 x 1, 87–88, 88*f*
  STM induced dissociation of, 36
Wide band gap semiconductors, 2–5
Wigner crystal, 62, 71
  formation of, 57–65
Wolkow, R, 79–81

Z

Zero-bias peak (ZBP)
  Kondo resonance and, 68–70, 68*f*
  in TTF-TCNQ compound on Au, 65
  vibronic coupling and, 69–70